ONE FOLD AT A TIME
ORIGAMI
38 EASY MODELS
FOR FIRST-TIME FOLDERS

PAUL JACKSON

TUTTLE Publishing

Tokyo | Rutland, Vermont | Singapore

CONTENTS

PLAY DESIGNS

GEOMETRICS

FROM NATURE

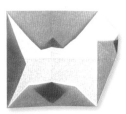

OBJECTS

Instant Box No. 1

Instant Box No. 2

Kevin's Boat

Sneaker

Arrow

Lark Box

Tent

Vehicles

Pajarita Envelope

Coffee Mug

Candle

Picture Frame

Strong Envelope

The Story of This Book

I created my first origami design when I was six years old, and the second design when I was fourteen. By the age of twenty-one, my designs numbered over a hundred. But, I was the odd one out among my creative origami friends in England. Almost everyone else was trying to engineer complex models, with increasing numbers of free points that would be used to create more limbs, more detail, more "Wow!" By contrast, my designs were simple, stylized, quick to make and easy to teach. I strove to achieve the minimum, whereas everyone else—apparently—was striving to achieve the maximum.

My simple approach wasn't dogmatic, it was just how my mind worked. Although I could fold complex designs from instructions, I couldn't figure out how to create designs that required an engineer's analytical brain. I worked intuitively, doodling with concentration, discovering models with unusual folding sequences, which logical analysis probably wouldn't produce. Gradually, I came to understand that for me, origami was a folding art, more than it was a folded art. It was ballet rather than sculpture.

As the years passed by, many of my creations were published and I wrote a number of books.

Then, in the early 2000s, I stopped creating models.

Why? Because parallel to my interest in creating origami models, I'd developed a career teaching folding techniques to art and design students. I began to focus on this aspect of folding and lost much of my interest in creating origami models, though I continued to enjoy folding other people's designs.

Then, COVID came. Lockdown. What does one do with all those hours at home? For some reason, I began to create models again; simple, concise, often unusual. The simplicity and enforced introspection of what I was creating reflected my lockdown lifestyle. Posting instructions on social media and teaching them live online, they found an audience, perhaps because they were quick to teach and easy to learn. Designs poured out of me in an unstoppable torrent. It was an extraordinary couple of years. Seemingly, almost every time I picked up a square of origami paper, I would create something I considered worth keeping, without any thought beforehand of what I might make.

The models in this book are some of those lockdown designs. They contain no advanced multi-fold manipulations. Everything is technically simple and thus, should be simple to make, though some have subtle nuances—simple technique does not necessarily mean simple folding.

I offer this book as a counterpoint to the many complex origami designs of recent years, which look amazing in social media posts and which rightly garner many "likes," but which are beyond the reach of most folders. Many of the models here can be folded in ten minutes or less. Hopefully, although simple, they offer original ideas, unusual sequences and a maximum of economy from a minimum of folding. They aspire to be the origami equivalent of a three-line, seventeen-syllable haiku poem. The designs also show creativity in free-flow, without direction, without a sustained technical theme and without preference for geometry, for living creatures … or whatever. The creativity is unfettered and spontaneous. If the book has a theme, it is "no theme."

Or rather, perhaps the theme is in the title, *One Fold at a Time Origami.* As you will discover, all the designs are made by making just one fold at a time, not making multiple folds simultaneously to form a reverse fold, squash fold, sink or any other of the more advanced techniques from which the majority of origami designs are created. In this sense, all the designs are made using the absolute simplest technical means. The origami term for this is "Pureland Origami," a term coined by the British origami expert John S. Smith in the 1970s, who himself created many Pureland designs.

The designs are presented to be enjoyed. The enjoyment can come from the act of folding, from sharing something you have folded, by teaching or displaying it, or simply from the satisfaction of completing a folding sequence. It is my hope that this book not only gives an insight into the mind of an origami artist, but that it also brings you many hours of pleasure.

— **Paul Jackson**

ORIGAMI SYMBOLS

If you have worked from origami instructions before, you will recognize that the standard set of symbols are used in the book. It is generally known as the Yoshizawa-Randlett-Harbin system of notation, devised and amended by three origami pioneers in the mid-1900s. The beauty of this system is that, used accurately, anybody in the world can work from a set of instructions, regardless of the language the accompanying text is written in and regardless of the language(s) understood by the reader. They make origami a universal language.

Like every author, I have introduced my own nuances, the most obvious of which is that I use three line weights, not two or even one, as other illustrators do. Existing creases are the thinnest weight, paper edges and folded edges are the middle weight and all the "to do" symbols are the heaviest weight. The reason for this is that the symbols are the most important feature of any step, and as such—in my opinion—they should be given prominence in the hierarchy of lines. I also think that three line weights give the steps more visual appeal and more depth on the page.

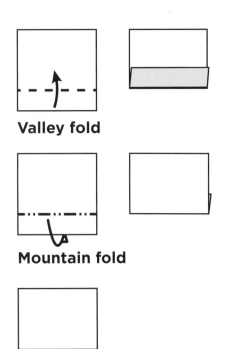

Valley fold

Mountain fold

Existing crease

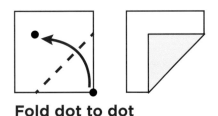

Fold dot to dot

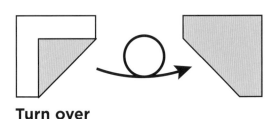

Turn over

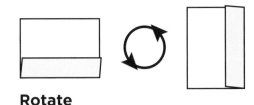

Rotate

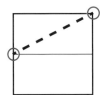

Note important locations

FOLDING TIPS

Origami Paper

Unless otherwise stated in the book, please use square origami paper. It is easy to find online or in art/craft stores, inexpensive and perfect for the job. It is lightweight and has different colors on the two sides, one of which is usually white. However, BEWARE of unscrupulous descriptions. Some colored photocopy and craft papers, even card weights, are sometimes described as "origami paper" or "suitable for origami," when in fact they are too thick and have the same color on both sides. Even if you buy from a retail outlet, beware. Shrink-wrapped packets can be described as "origami paper" and look authentic, but it can be difficult to see if the paper is thin and has different colors on the two surfaces. So the rule of thumb is: before you purchase online or from a store, be sure your paper is suitable. Check the descriptions carefully. If in doubt, don't buy.

Your Folding Space

Folding paper should be an enjoyable experience. Taking a few moments to set up a suitable workspace will increase your pleasure and make you want to fold for longer periods of time.

Fold on a flat, hard surface, such as a table. If the table is overflowing with clutter, clear it away so your area is tidy, open and you can breathe.

Don't fold with a light placed directly overhead. This will flatten the shadows on your paper and make folding very difficult. Instead, fold with a light low-down and to the front of you. This can be a window, an artificial light such as a desk lamp or even a ceiling light. However, the soft light from a window is much more beautiful than any artificial light. So, perhaps reposition a chair, move a work surface or play with the lighting arrangement in a room until you have created a setup in which folding is a pleasure. It really WILL make a difference.

If at First You Don't Succeed...

...go and do something else, and then come back and try again! Folding instructions are a kind of puzzle—codes to be cracked. Sometimes a solution can be elusive, but with patience and a cheerful but determined outlook, you will succeed. Of course, the more you fold, the easier it becomes!

The Paper is Your Partner, Not your Enemy

Treat your paper with respect. Touch it lightly and carefully. Don't hurt it. Don't whack it, bash it, thump it, smash it, injure it or mistreat it in any way. The paper is your partner, not an adversary to be conquered. Experienced paper folders sometimes talk about "the paper ballet," referring to a kind of pas-de-deux between your hands and the paper, as you fold. If you fold elegantly, deliberately, with care and with love, you will enjoy the experience more and your work will look better.

HOW TO USE THIS BOOK

In some ways, an origami book is an origami book is an origami book. They are all broadly similar and they all have their distinctive features. This book differs from most because the pace of its instructions is relatively slow. Here's why.

Many books, especially those containing more advanced designs, will assume a level of technical proficiency from its readers, so the instructions become dense and busy. A lot happens on each step. However, if the designs are so simple that they can be attempted by a beginner, this compression of many steps into a few drawings can be very confusing. It is better to give each new fold and each new instruction—such as "turn over"—its own step. This means that a design that takes eight steps to describe with busy compound drawings might take twice as many steps to describe in an expanded fashion. However, the expanded sixteen steps will be easier to follow than the eight aforementioned steps, even though they contain the same instructions.

Also, the diagrams in this book are larger than might have been expected. Why? Because for a beginner, larger drawings are easier to follow than smaller ones. They are less cramped and have room to breathe, making them less stressful to view and easier to understand.

The instructions in this book contain a minimum of text. Most steps are very self-evident. The constant repetition of "fold in half" or a similar phrase, is unnecessary. Instructions are given only when something unusual or requiring precision happens.

How do I know that relaxed, expanded folding sequences, larger diagrams and only

essential text are helpful to beginners? My wife, the origami artist and educator, Miri Golan, and I have an EdTech start-up that teaches the full K–7 geometry curriculum, using origami as a learning tool. We have tested, tested and tested different ways to teach, so that everyone in a class can understand the instructions. We found that if the instructions were slow-paced and large, almost everyone could follow them successfully.

So, this book follows that same method of presentation. The style is a little unorthodox, but because the priority of any book should be to help the reader fold with success, I think it's well worth sacrificing a few extra designs for extra clarity. I hope you agree.

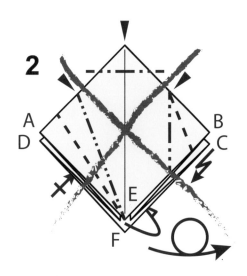

WHY FOLD ORIGAMI?

This might seem like a strange question considering that you're partway through reading this book, but it's one worth reflecting upon. Why fold? For what purpose? What, if anything, do we gain from it?

Folding paper is unique in at least one way. Usually when we make something, we hold a tool such as a hammer, needle or a pencil, or our hands operate machinery such as a sewing machine, bandsaw or a mouse. We almost never use our hands to make something directly; it's the manipulated proxy tool that does the making. However, with origami, there are no intervening tools or equipment, we make something ONLY with our hands. WE do the work. We have direct contact with the material and we are in sole control of everything. There is no proxy tool.

This direct contact with the material means that folding paper is an almost unique creative experience. It has a power and a personal significance beyond the making of objects using tools, creating a heightened sense of achievement and satisfaction.

Further, the transformation of a sheet of ordinary paper into something cool, beautiful, fun or ingenious is almost alchemical. We are conjuring something significant from something ephemeral (paper). We are magicians ... except that origami isn't stage magic but true magic; there are no illusions or deceptions. This process of transformation is immensely satisfying to experience, whatever our age, gender or origin. The joy of origami is universal.

Folding paper is fun, to be sure, but it is important to note that it is also a profound activity that speaks both to the emotions and to the intellect, ultimately improving our sense of well-being. Folding paper is good for us!

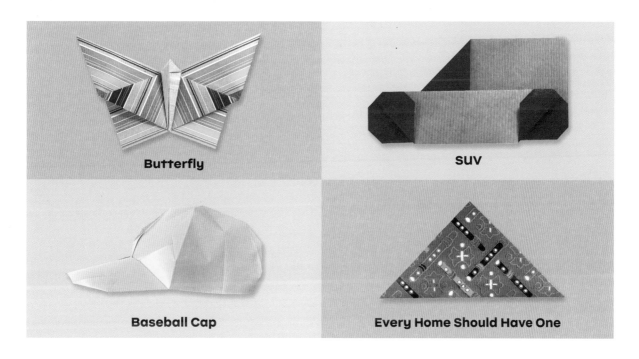

Butterfly

SUV

Baseball Cap

Every Home Should Have One

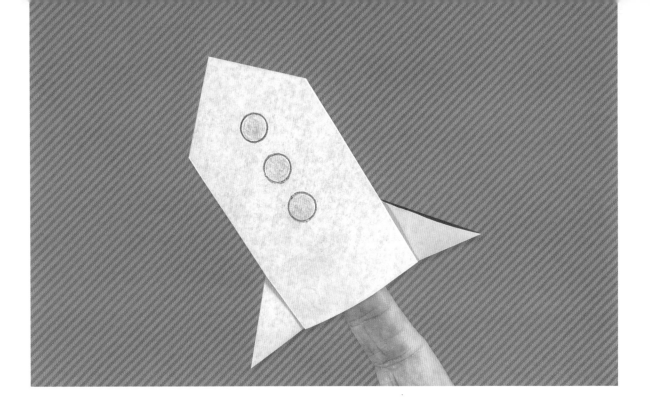

Finger Rocket

Origami finger puppets are easy to play with because they have no moving parts and stay securely on the hand. Typically, the puppets are animals or faces, but this one is unusual, being a rocket. Add drawn-on windows, stickers and more to the final model, to customize what you fold. What is your mission? Where are you going? With whom?

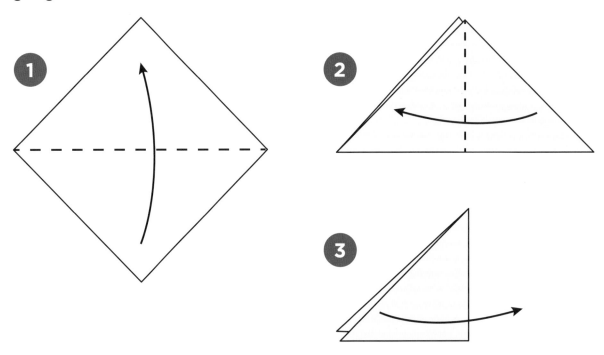

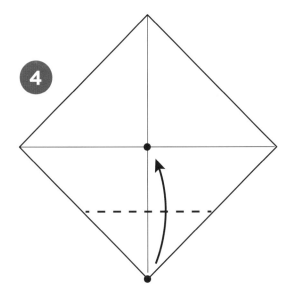

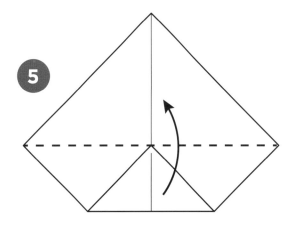

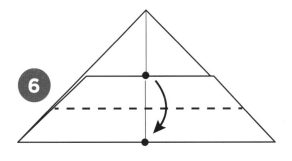

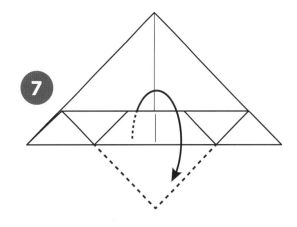

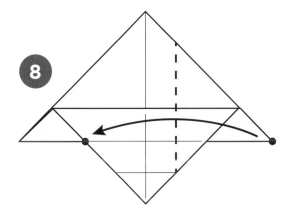

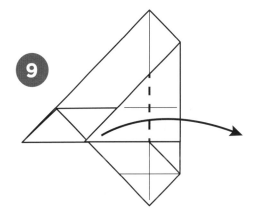

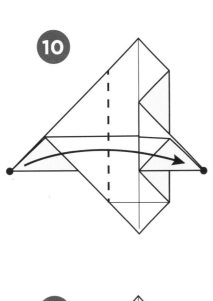

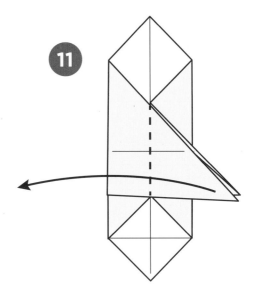

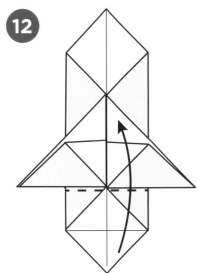

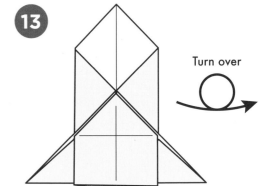

Turn over

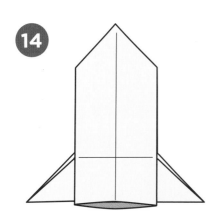

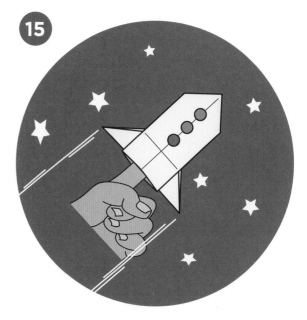

Insert a finger between the layers at the base of the rocket.

Talking Toy

There are many origami action toys in which a mouth is made to move. However, I think this design is unique, because like a real mouth, the upper part doesn't move, while the lower part—the lower jaw—moves up and down. In all other origami designs, the two halves of the mouth move equally, which looks rather strange! You will need to hold the paper exactly as instructed to move only the lower jaw.

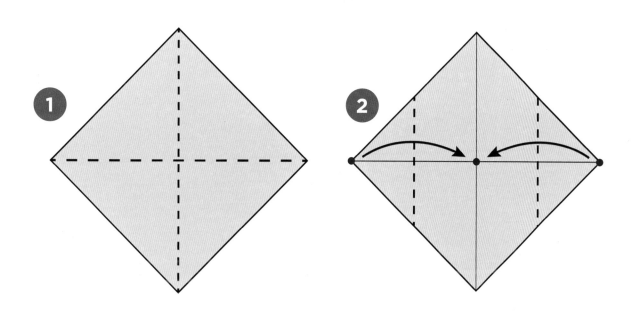

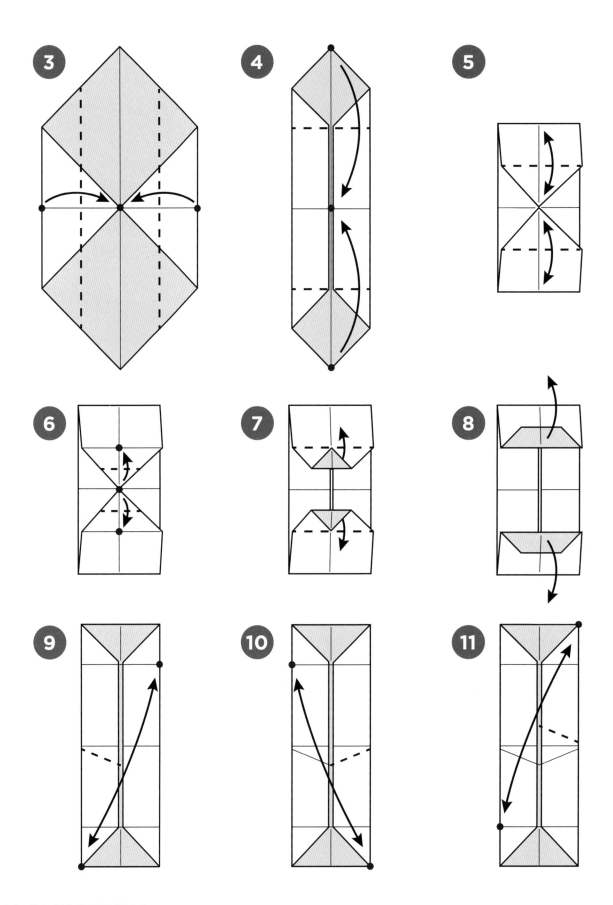

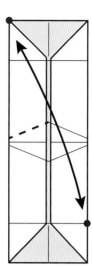

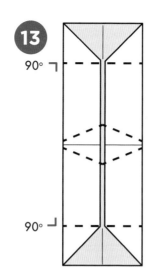

90°

90°

Sharpen the four creases as shown, so that the paper becomes 3-D.

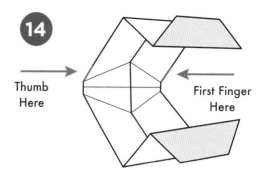

Thumb Here

First Finger Here

This is the key step. When the paper is 3-D, hold the edges marked in red between your thumb and first finger. Note that these edges are IMMEDIATELY ABOVE the "V" folds that stretch across the paper. Do not hold the short sections of edge between the "Vs."

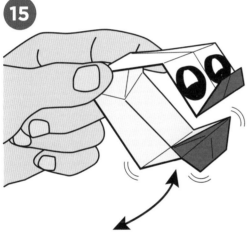

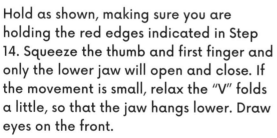

Hold as shown, making sure you are holding the red edges indicated in Step 14. Squeeze the thumb and first finger and only the lower jaw will open and close. If the movement is small, relax the "V" folds a little, so that the jaw hangs lower. Draw eyes on the front.

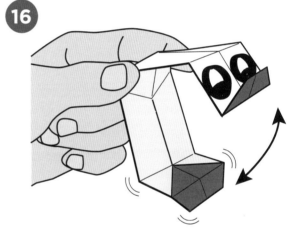

The reason the lower jaw moves while the upper jaw remains stationary is because of the way the finger and thumb are positioned on the edges of the paper. Almost ANY design with the same configuration of "V" fold creases in the middle of the paper will "talk" in the same way, moving the lower jaw, but not the upper jaw. In making this toy, I created probably fifteen different variations, wider, narrower, longer, shorter, with and without color-change eyes, before deciding this version was the best. See if you can create your own version!

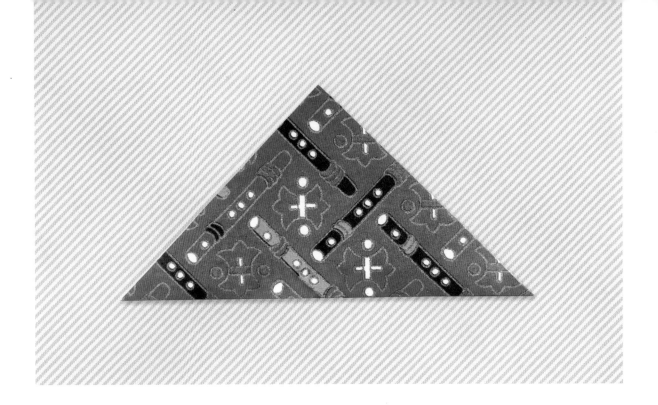

Every Home Should Have One

Yes ... this design is a joke, but one with a point. A flat, locked triangle is as dull and as pointless as an origami design can be, right? And yet, the locking mechanism is simple and interesting, so the design has some technical merit, even if it looks boring. So, what is a model? What is worth folding and why? Is origami a folding or a folded art? It's a design that asks questions. Do you have answers?

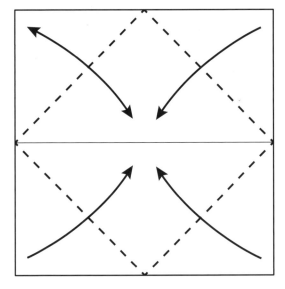

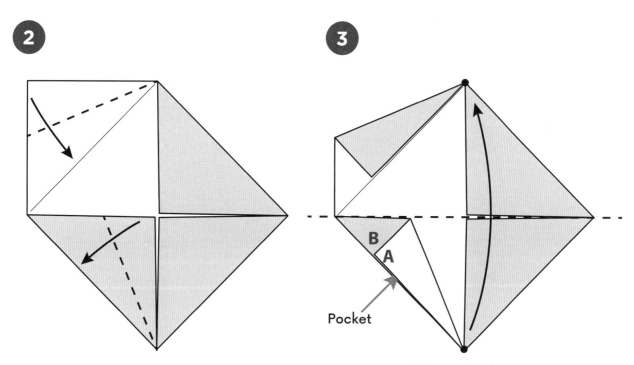

Note A, B and the pocket behind A.

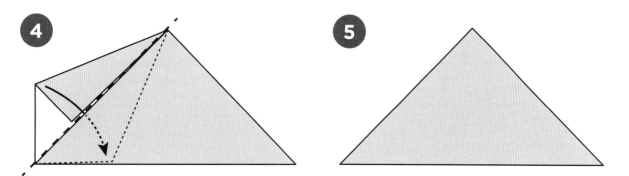

Tuck in the triangle so that it slots into the pocket between A and B, holding the paper tightly shut. Press the paper very flat, sharpening all the creases.

Bird Beak

Two very simple units are subtly interlocked so that each holds the other in place, creating a three-dimensional beak. The small triangle held by the fingers is particularly important, because it keeps the assembly stable. Sometimes, it's the smallest detail that holds the key to a design. Is it possible to interlock the units in a different way, to create a color-change between the top and bottom halves of the beak?

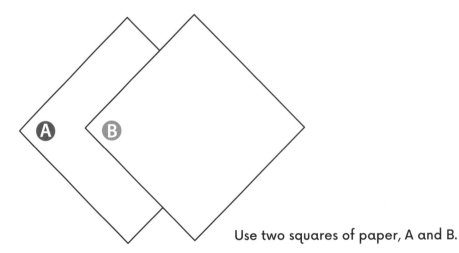

Use two squares of paper, A and B.

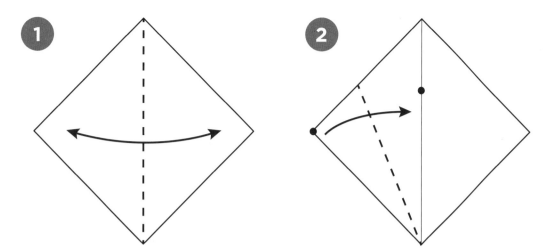

Fold A and B the same, up to Step 4.

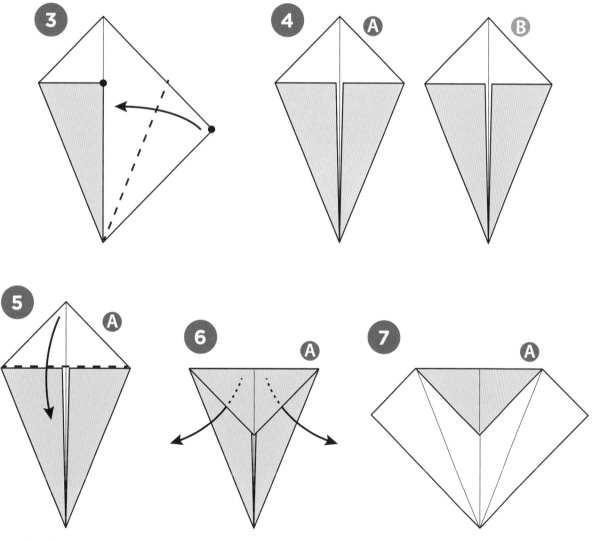

Fold only square A, as shown.

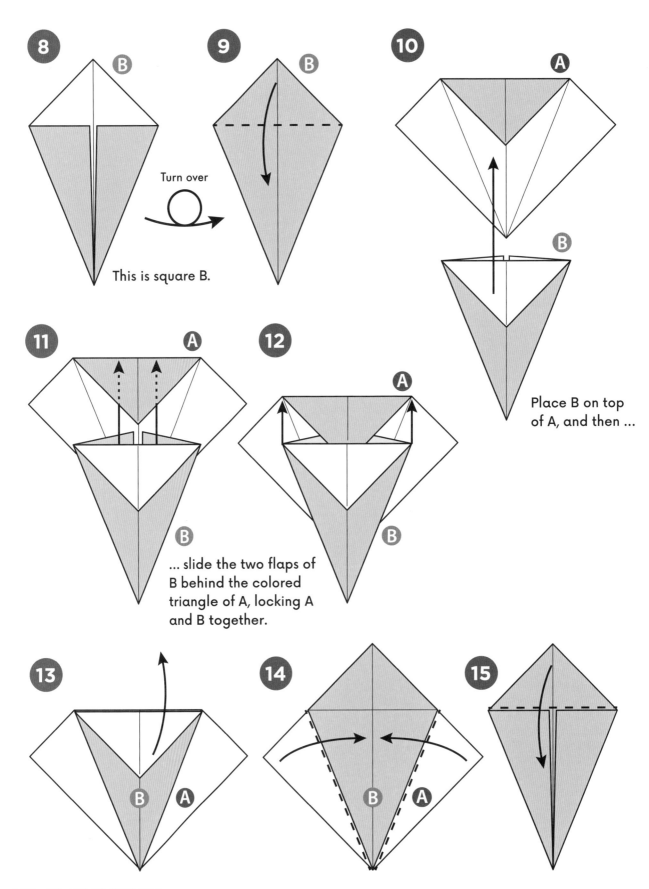

8

B

Turn over

This is square B.

9

B

10

A

B

Place B on top of A, and then ...

11

A

B

... slide the two flaps of B behind the colored triangle of A, locking A and B together.

12

A

B

13

B A

14

B A

15

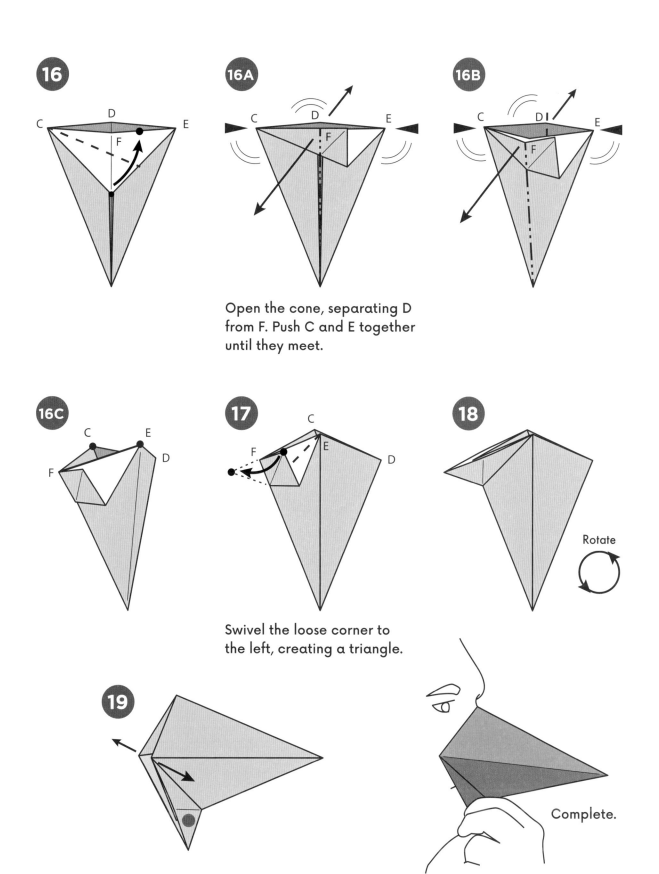

16

C　　D　　E
F

16A

C　D　E
F

Open the cone, separating D from F. Push C and E together until they meet.

16B

C　D　E
F

16C

C　E
F　　D

17

C
F　E
　　　D

Swivel the loose corner to the left, creating a triangle.

18

Rotate

19

Complete.

Angry Bird

Just small movements of the fingers will make the bird swing wildly left and right. Make two and have a battle royale with a friend. When not in combat, the design will stand well. The rubbing of the fingers, as though sprinkling salt, is an underused method of powering an origami design. Can you create a toy such as a fish or a mouse that swishes from side to side, powered by "sprinkling salt?"

1

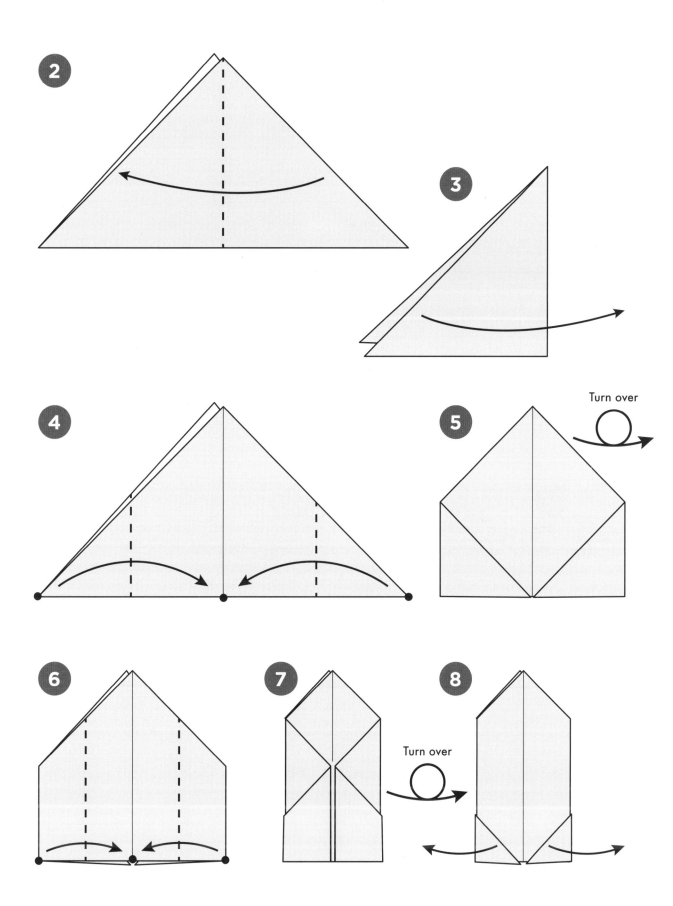

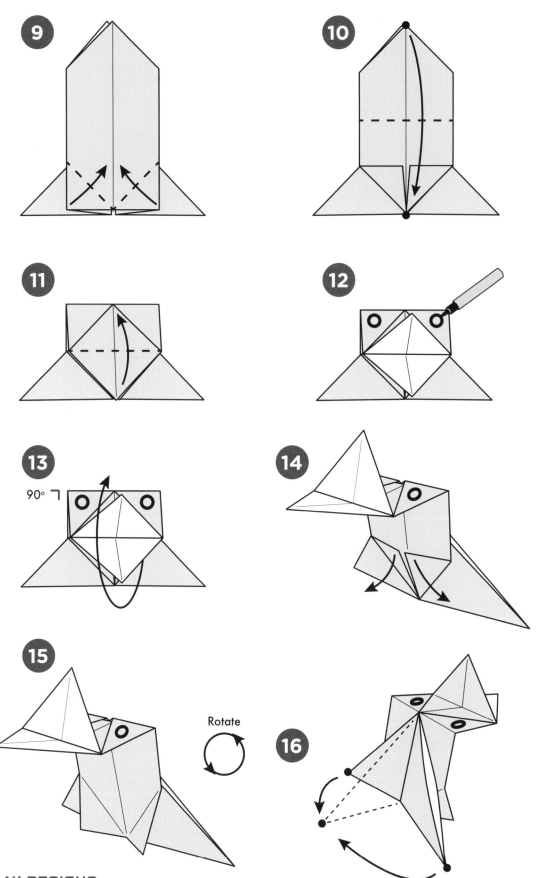

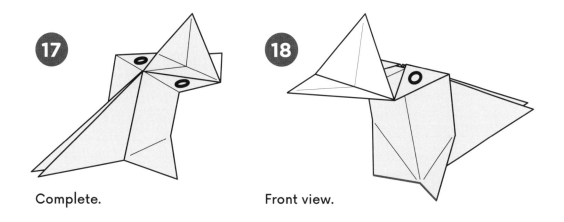

17 Complete.

18 Front view.

19 SQUAWK!! SQUAWK!!

The bird will stand.

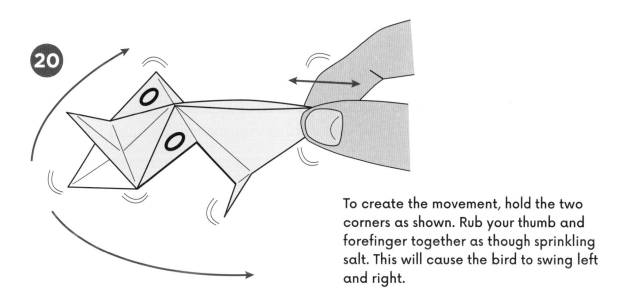

20 To create the movement, hold the two corners as shown. Rub your thumb and forefinger together as though sprinkling salt. This will cause the bird to swing left and right.

Baseball Cap

Use Tabloid size paper in the US or A3 elsewhere. If these papers are unavailable, tape together two sheets of 8½ × 11 inch paper or two sheets of A4 paper. Unlike many origami hats and caps, this one fits snugly on the head and will remain there all day, if you're so inclined. Be sure to make the curved fold at the end with extra care. This is the fold that makes the visor stay horizontal.

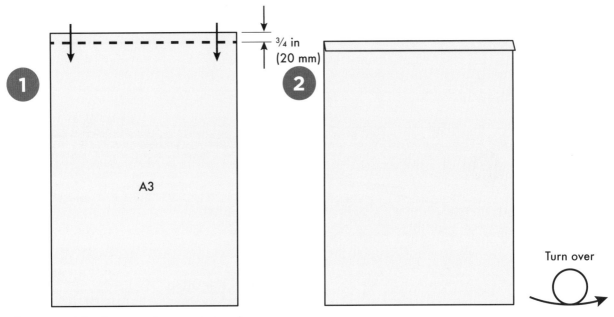

If using Tabloid paper (or two joined Letter-size sheets), trim a 1½ inch strip from the top edge of the paper first.

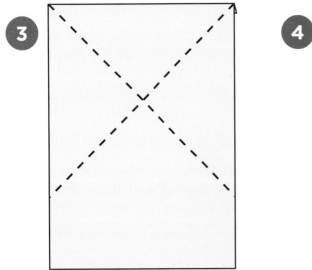

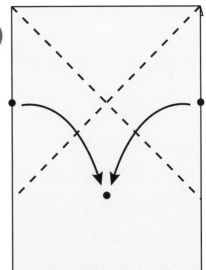

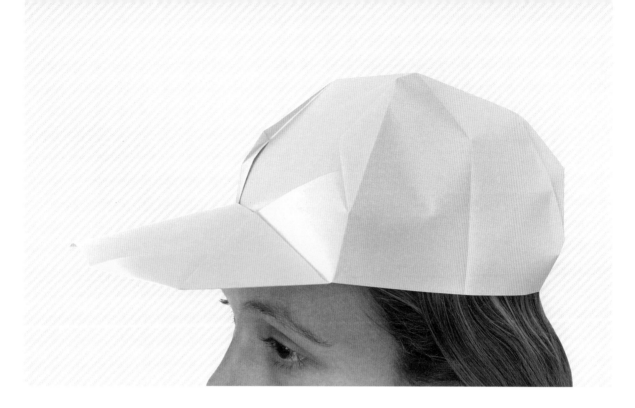

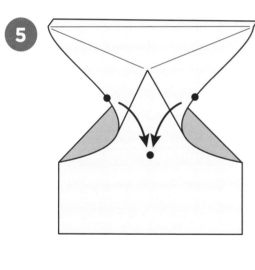

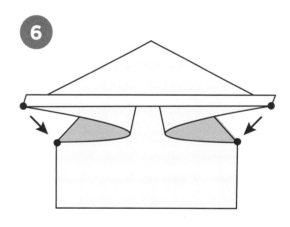

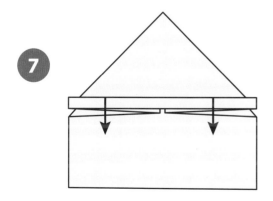

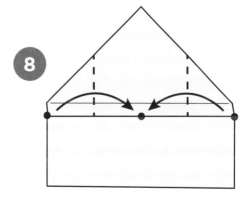

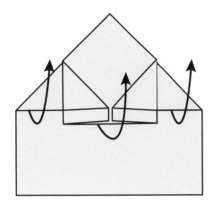

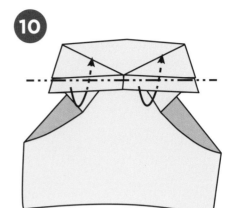

Fold the edge inside, using the fold made in Step 1.

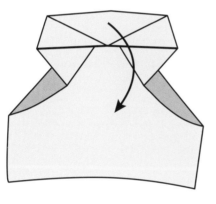

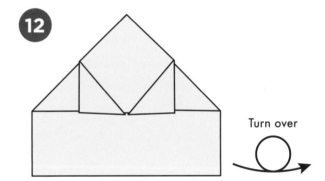

Turn over

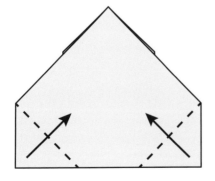

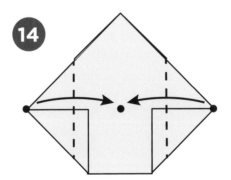

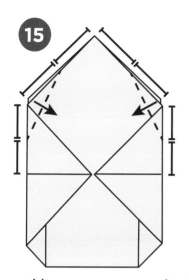

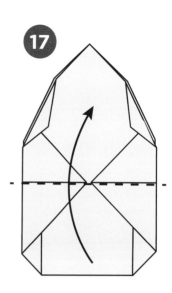

Fold two narrow triangles. Try to make the folds end at the mid-points of the edges, as shown.

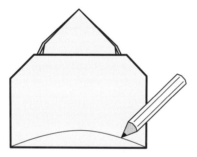

Draw a curve, the top of which is about ¾ in (2 cm) from the bottom edge of the paper.

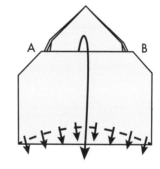

This is the step that's important to do well. Crease the curve a little as a time, following the pencil line. This will bring A and B down to the horizontal and create a curved visor, locked into position.

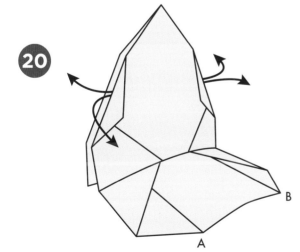

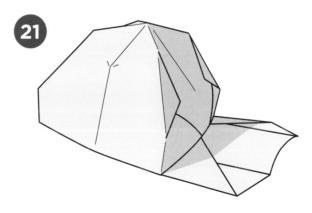

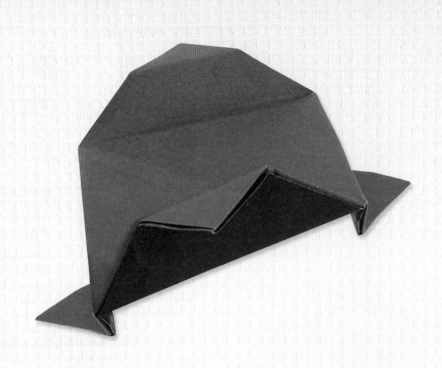

Jumping Jacks

The design is essentially a square with a two-fold zig-zag pleat across each corner. The thicker the square, the better the spring, so instead of origami paper, you could try using Bristol board or a cereal box. Does it make a difference to the spring if you make a square from many layers of paper, or from one thick layer of card? What is the highest and farthest you can make a Jack jump?

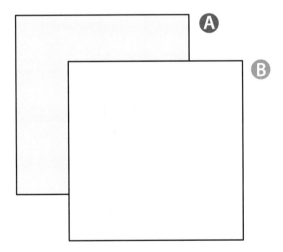

Use two squares of paper, A and B.

Folding a Unit

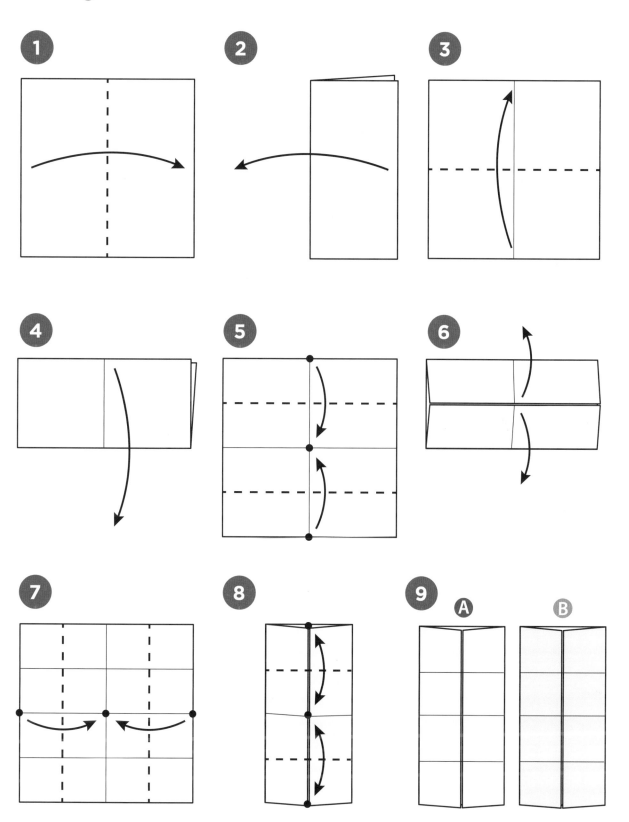

Assembling the Units

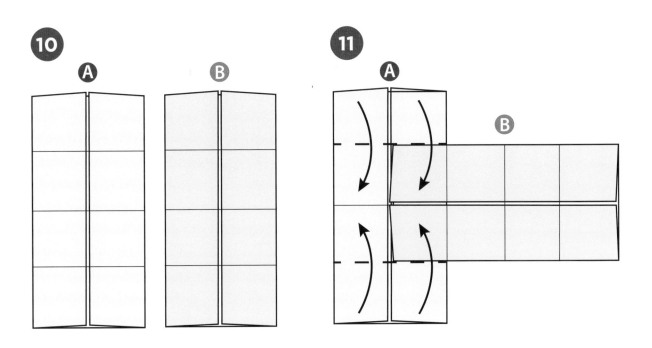

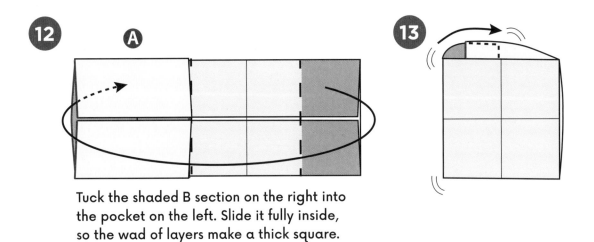

Tuck the shaded B section on the right into the pocket on the left. Slide it fully inside, so the wad of layers make a thick square.

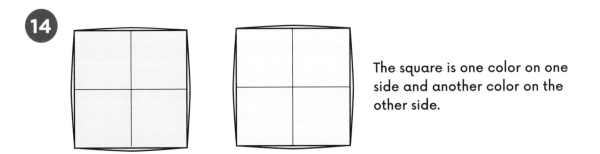

The square is one color on one side and another color on the other side.

Creating the Springs

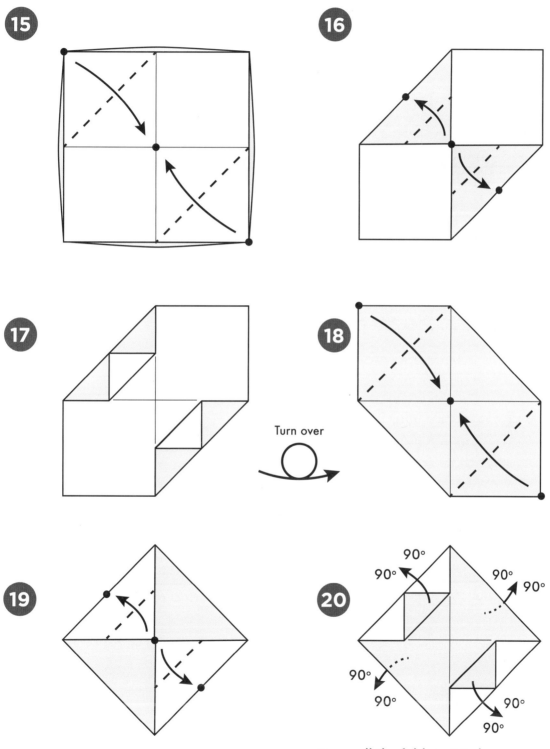

15

16

17

Turn over

18

19

20

90° 90° 90° 90°

90° 90° 90° 90°

Open all the folds to 90 degrees.

21

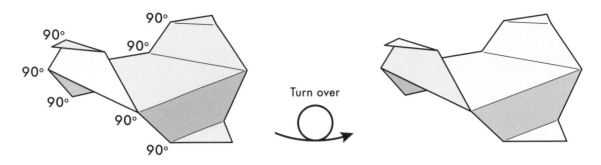

Turn over

22

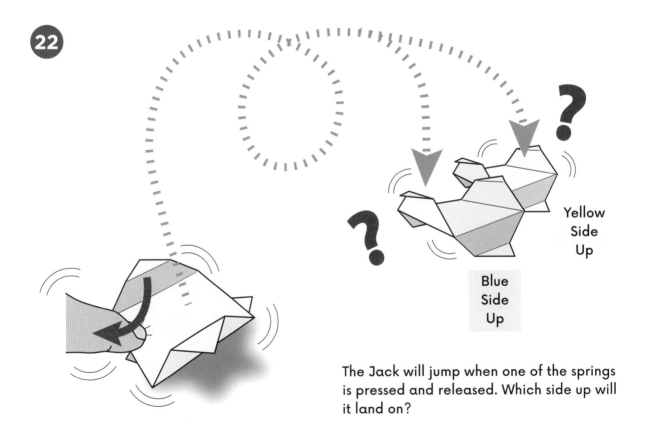

Blue
Side
Up

Yellow
Side
Up

The Jack will jump when one of the springs is pressed and released. Which side up will it land on?

Variations

The Jack is a square, with extra folds across each corner to create the springs. So, any folded square can become a Jack, providing there are enough layers to create good springs.

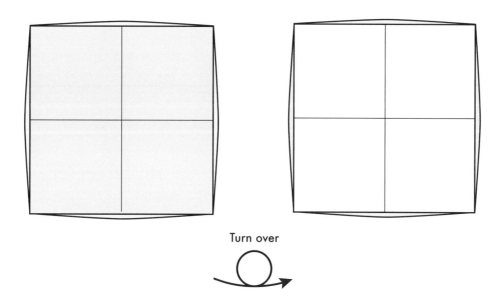

Turn over

One Square of Origami Paper

Fold the corners to the center point two or three times to create a small square with many layers.

One-Sided Springs

Fold all four springs onto the underside of the square. If the Jack jumps and lands on its spring-less back, you lose!

Coloration

Use a square of origami paper and think how you might show the colored side and the white side on the two faces of the square.

Other Materials

The Jack can also be made from thicker papers, such as copy paper or drawing paper. They can also be made from cereal boxes or similar thin carton/duplex paperboard/Bristol board.

Games

There are many games that can be played using one or more Jacks, with others or by yourself. Here are some suggestions.

Moji-Jacks

Draw happy and sad faces on opposite sides of the finished Jack.

Score ten points if a jumping Jack lands on the happy side and five points if it lands on the sad side. Challenge friends to see who can reach 100 points first, or time yourself.

Left-Right Flip-Flop

Flip the Jack using alternate hands: first one hand, then the other, then the first, then the other again ... and so on.

Flip and Catch

Jump a Jack and before it lands, try to catch it cleanly. It's more difficult than you might think! Try alternating your flipping and catching hands.

Jack Crazy Golf

The Jacks have a mind of their own and rarely jump where you want them to jump. So, challenge friends to a game of Crazy Golf. Each player tries to jump his or her Jack into a small container, positioned centrally between the players. Turns are taken, as in golf.

The distance to the "hole" (the container) can be very small, 1 foot (30 cm) is sufficient.

Jack Olympics

Try to design Jacks that will jump high, jump a long way or be ultra-acrobatic.

What size should they be?

Should the folding be modified?

What material(s) should you use?

Twice-as-Big Sonobe

The Sonobe Unit is perhaps the most famous unit in all origami. First created in the 1960s, it has spawned many variations. This variation is perhaps unique because it makes a standard unit, but it is twice the area of the original. The reason is the introduction of a cut—heresy to many origami purists, but sometimes bending the rules can yield unexpected and worthwhile results. Is it a cheat?

To make the example shown here, you need twelve identical squares.

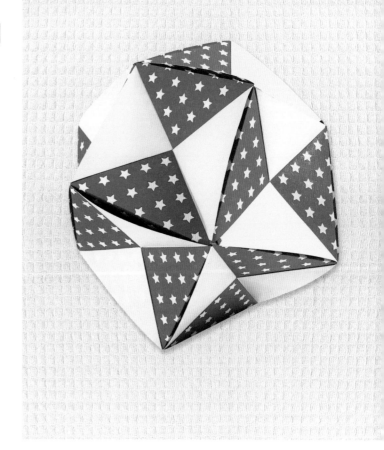

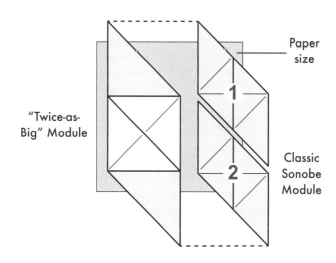

"Twice-as-Big" Module

Paper size

Classic Sonobe Module

1

2

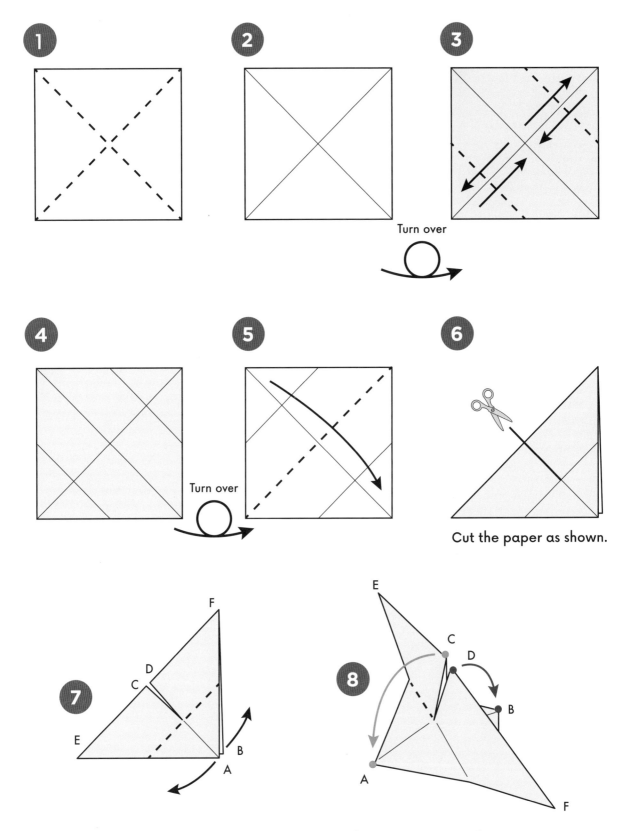

1

2

3

Turn over

4

5

Turn over

6

Cut the paper as shown.

7

8

Flatten C onto A and D onto B.

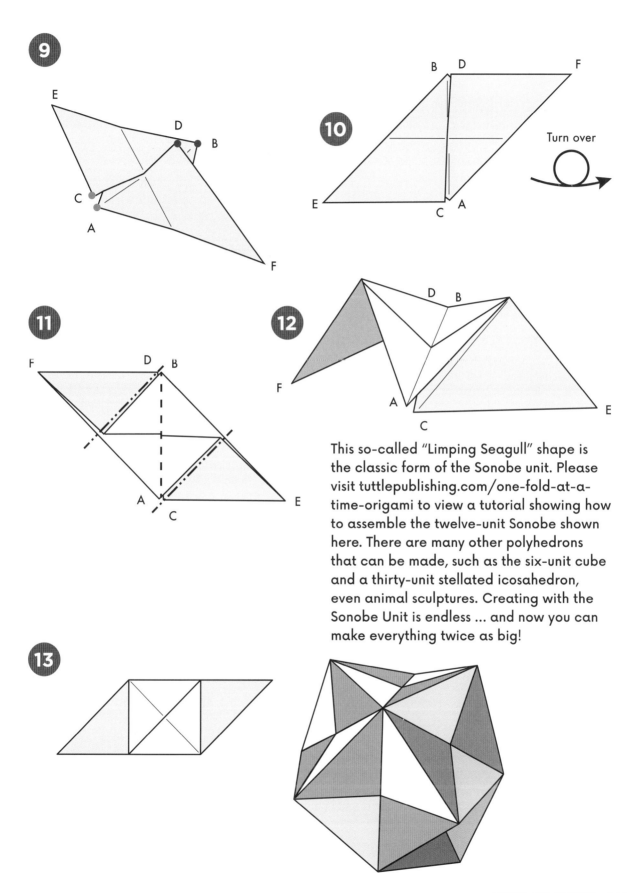

9

10 Turn over

11

12 This so-called "Limping Seagull" shape is the classic form of the Sonobe unit. Please visit tuttlepublishing.com/one-fold-at-a-time-origami to view a tutorial showing how to assemble the twelve-unit Sonobe shown here. There are many other polyhedrons that can be made, such as the six-unit cube and a thirty-unit stellated icosahedron, even animal sculptures. Creating with the Sonobe Unit is endless ... and now you can make everything twice as big!

13

Tetrahedron

The 8½ × 11-inch Letter paper used in North America has no special mathematical proportion. However, it does have at least one coincidental geometric quirk, exploited here: if its width is that of an equilateral triangle, the height is one and a half equilateral triangles, at least to within an error of 0.03%. To make a tetrahedron from the crease pattern becomes a simple matter.

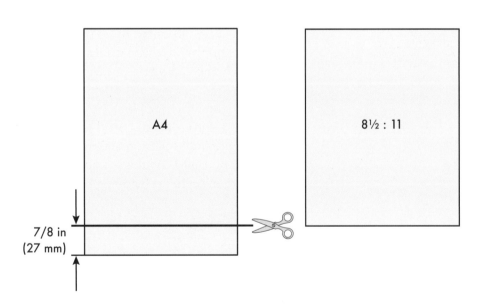

A4

8½ : 11

7/8 in
(27 mm)

If you are using A4 paper, trim 7/8 of an inch (27 mm) from the bottom edge. This will not create the dimensions of 8½ × 11 inches, but it will create a slightly smaller rectangle with the same proportions. If you are using 8½ × 11-inch paper, you do not need to do anything and are good to go!

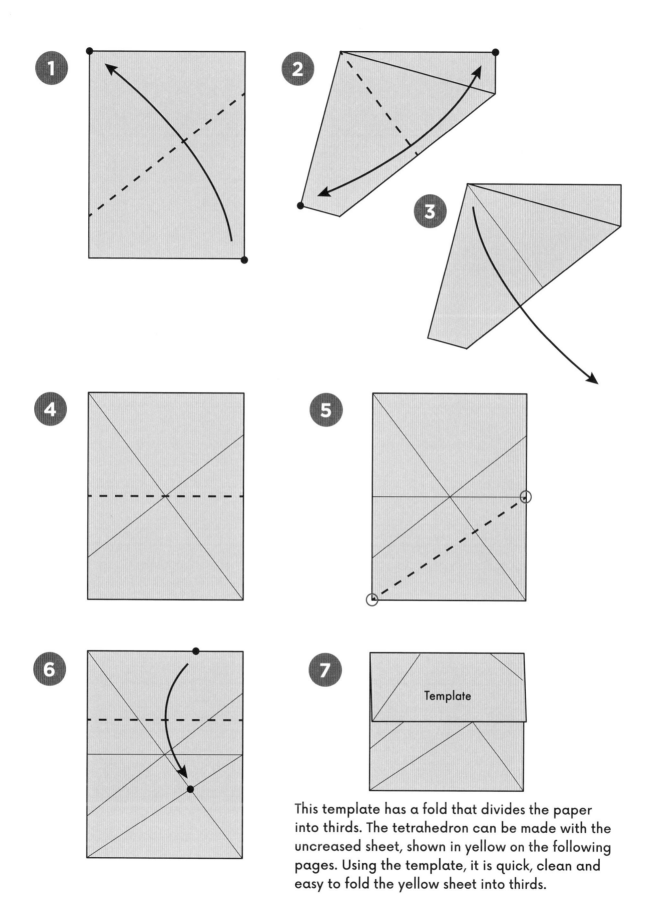

This template has a fold that divides the paper into thirds. The tetrahedron can be made with the uncreased sheet, shown in yellow on the following pages. Using the template, it is quick, clean and easy to fold the yellow sheet into thirds.

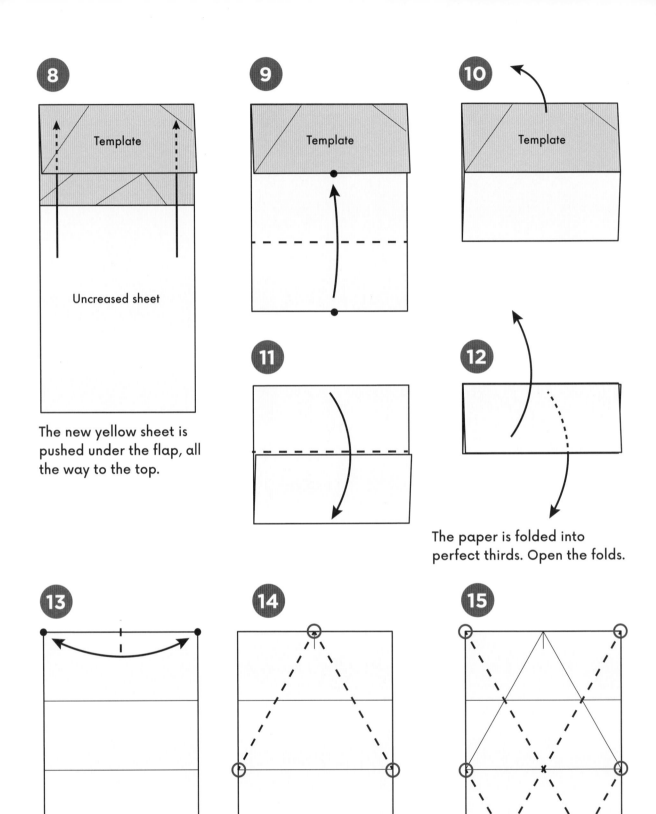

8

Template

Uncreased sheet

The new yellow sheet is pushed under the flap, all the way to the top.

9

Template

10

Template

11

12

The paper is folded into perfect thirds. Open the folds.

13

14

15

Complete the grid, taking great care to fold accurately.

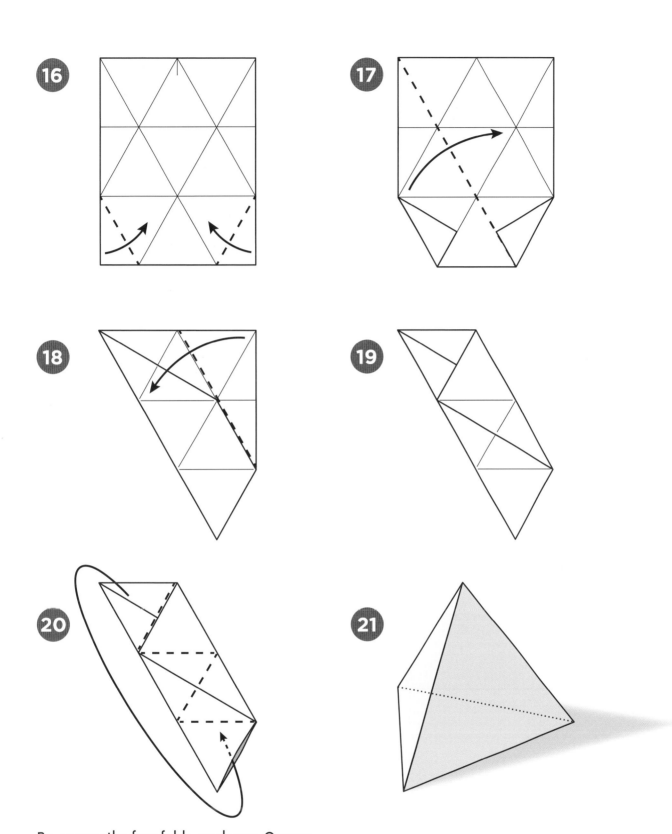

Re-crease the four folds, as shown. Crease them strongly. Create the tetrahedron by tucking the top corner into the pocket along the bottom edge.

Two-Sheet Version

Four-Sheet Version

Mirror Collages

I discovered this curious effect many years ago, but only in recent years have I explored it deeply. By folding mirror image and opposite-color units, the curious illusion is created of squares passing through each other, apparently in the same plane. Can you see what is happening? It really is a very strange phenomenon! Even stranger, is that the fold(s) can be placed anywhere, providing a few simple rules are followed.

Two-Sheet Version

1

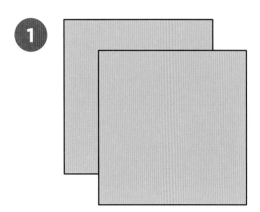

Begin with two identical squares of origami paper.

2

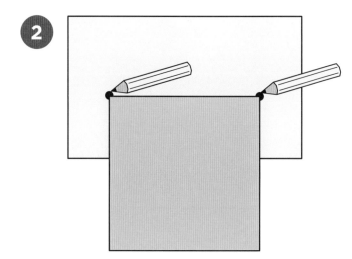

Take a piece of scrap paper and mark on it the position of the top corners of the origami paper square.

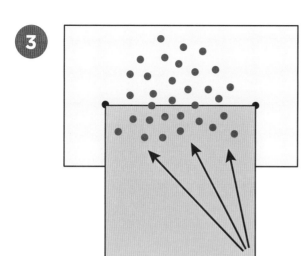

3 Bring the bottom corner of the square up to any dot in the cloud of dots, and then fold.

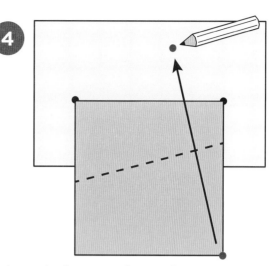

4 I chose the location shown above. Your choice may look different, and that's okay.

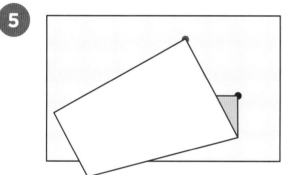

5 This is my result, but yours may look different.

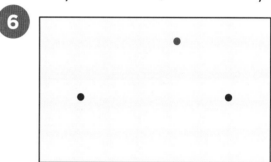

6 The scrap paper will look like this. Keep it as a template.

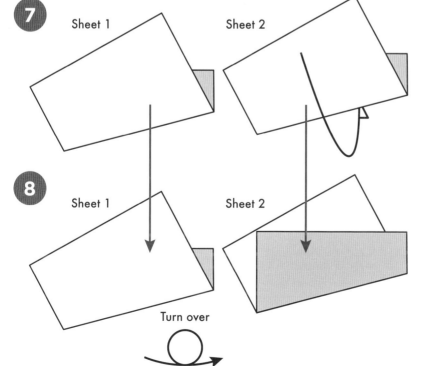

Sheet 1   Sheet 2

Fold the Sheet 2 in exactly the same way, using the template as a guide. Once made, reverse to fold to the rear side.

Sheet 1   Sheet 2

Now, turn Sheet 1 over.

Turn over

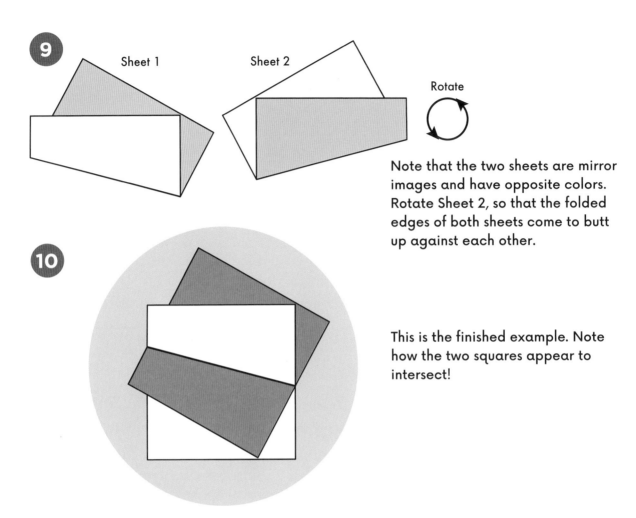

9

Sheet 1   Sheet 2

Rotate

Note that the two sheets are mirror images and have opposite colors. Rotate Sheet 2, so that the folded edges of both sheets come to butt up against each other.

10

This is the finished example. Note how the two squares appear to intersect!

Four-Sheet Version

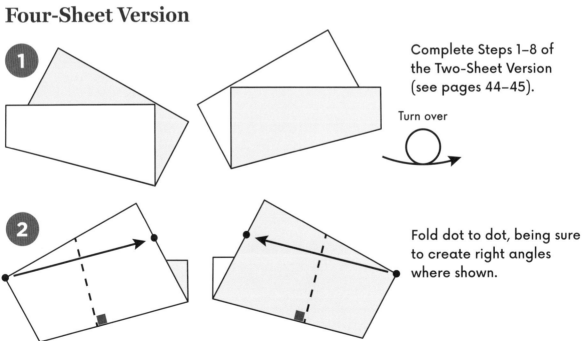

1

Complete Steps 1–8 of the Two-Sheet Version (see pages 44–45).

Turn over

2

Fold dot to dot, being sure to create right angles where shown.

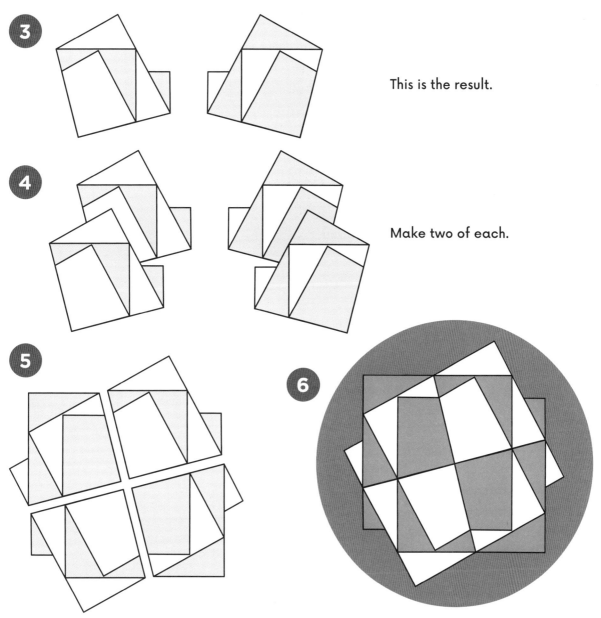

3 This is the result.

4 Make two of each.

5 Arrange them as shown and join them together.

6 The effect resembles two colored squares and two white squares, apparently intersecting many times. Can you find them? It's a complex and fascinating effect, achieved by the simplest of folding.

Further Ideas
- What other illusions can you create using this method?
- Can you make an illusion from six or eight squares, with angles of 60 or 45 degrees at the center point?
- What happens if you use rectangles, or even triangles or circles?
- Does the effect work if the collages are multi-colored?
- Can you create the effect, but in an asymmetrical arrangement of the units?

Two-Tone, Two-Piece Ornament

This decoration has six blades that radiate from the center. The top half is one color, the lower half is another. Contrary to expectations, the two units are not interlocked from the top and bottom, but from the left and right! This highly unorthodox method of construction means that a complex form can be achieved with simple folding. Be sure to follow the lettered corners with great care.
Use two squares, A and B.

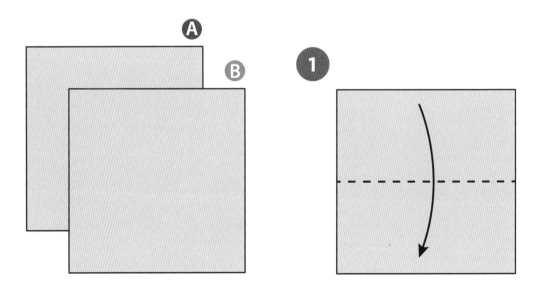

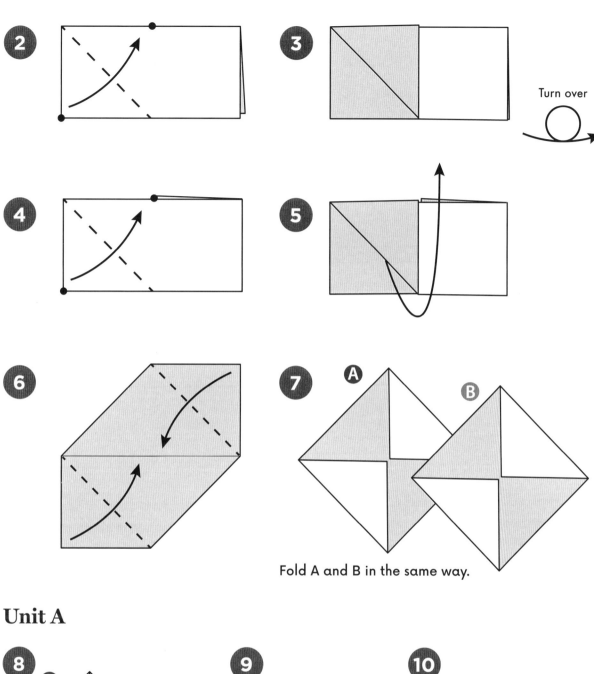

Fold A and B in the same way.

Unit A

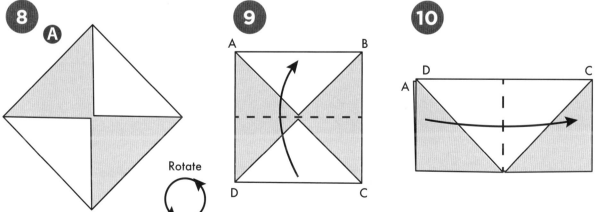

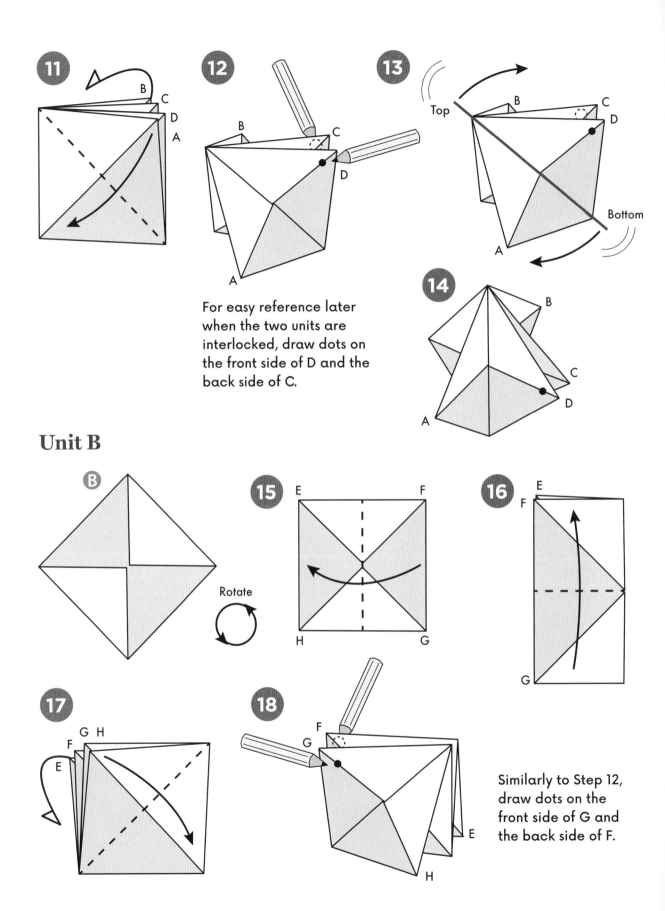

For easy reference later when the two units are interlocked, draw dots on the front side of D and the back side of C.

Unit B

Rotate

Similarly to Step 12, draw dots on the front side of G and the back side of F.

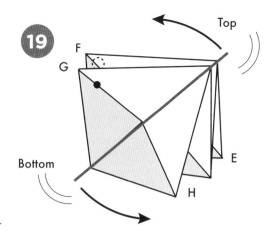

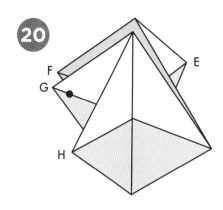

Assembly

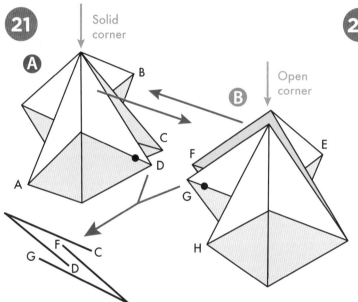

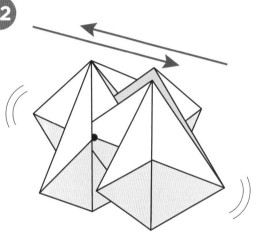

Continue to interlock the units.

Hold the two units as shown, so that corners D and C face corners G and F. Notice also that unit A has a solid corner at the top, whereas unit B has an open corner at the top. Arrange D, C, G and F as shown, in a right-left-right-left pattern. This will lock the two units together.

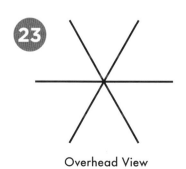

Overhead View

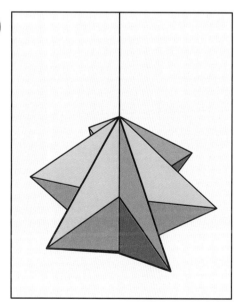

Two-Tone Decoration

There are a great number of four-bladed decorations created from Preliminary and Waterbomb Bases, made by interlocking two units. This decoration looks like many of them, but is made from just one square of paper. Somewhat bizarrely, the paper is pleated into sixths to achieve the effect. In origami, as with most things in life, there are often multiple ways to achieve the same result, some more unexpected than others.

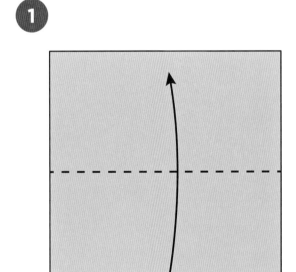

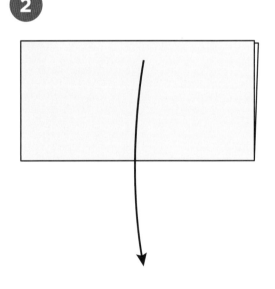

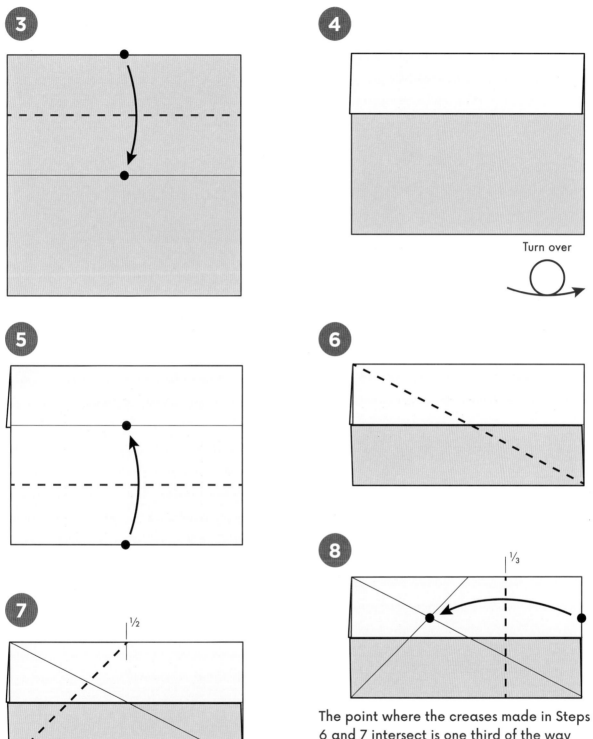

Turn over

Pinch the halfway point along the top edge, and then crease as shown.

The point where the creases made in Steps 6 and 7 intersect is one third of the way across the paper. When the opposite edge is folded to this intersection, the fold made will likewise be one third of the way across the paper. This is the beginning of folding the paper into sixths.

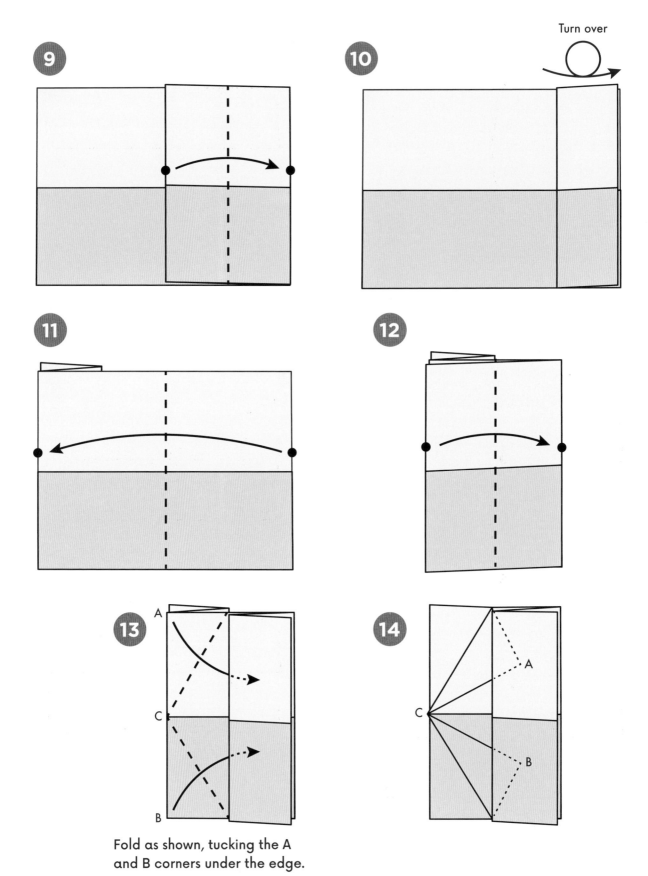

Turn over

Fold as shown, tucking the A
and B corners under the edge.

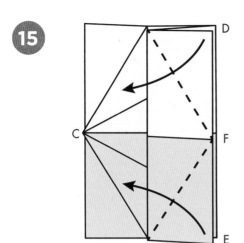

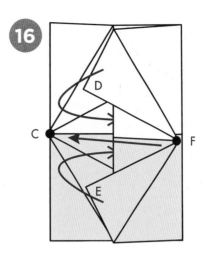

Fold F across to lie on top of C. At the same time, rotate D and E inside the right-hand pocket, to lie on top of A and B.

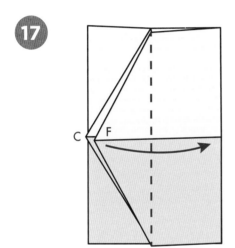

Fold F back to the right, leaving corners D and E inside the pocket.

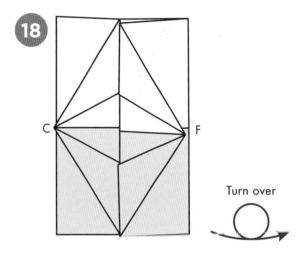

Turn over

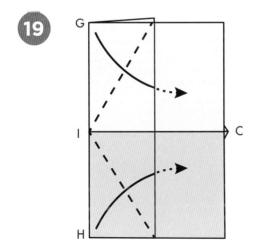

Repeat Step 13.

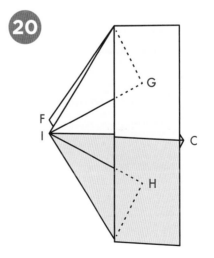

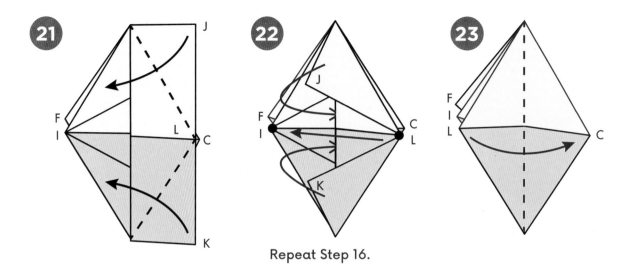

Repeat Step 16.

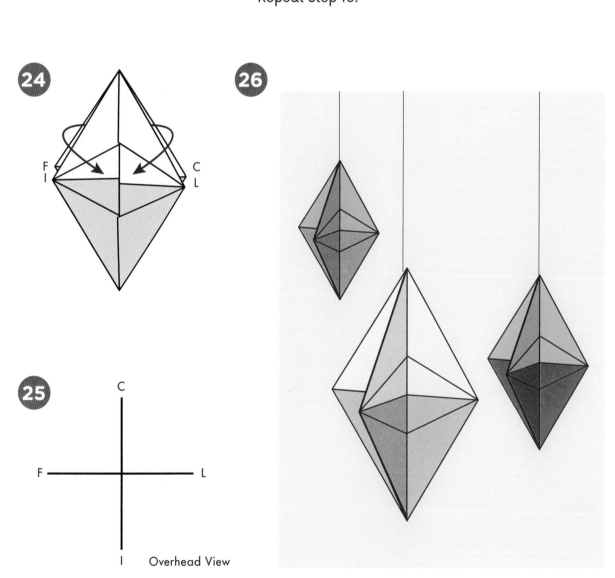

Overhead View

Three-Dimensional Star

This eight-piece unit design is unusual because, unlike most modular structures, it doesn't lock together by having loose flaps slid into pockets. Instead, it holds its shape by having opposing flaps wrap around a common triangle. When the eighth unit is interlocked, each unit presses equally against its neighbors and the structure is stabilized.

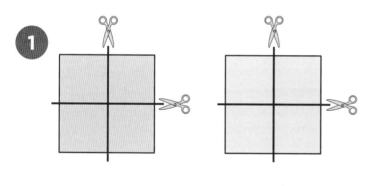

Take two squares of differently colored 6-inch (15-cm) origami paper and cut each into quarters. All eight squares will be folded in the same way.

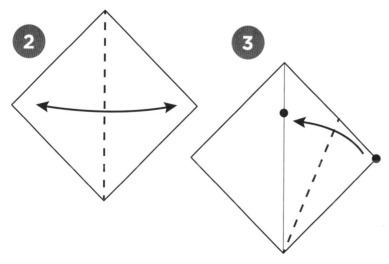

Note that it is the RIGHT HAND corner that is being folded to the center line and the crease terminates at the BOTTOM corner. This "handedness" must be repeated exactly on all units, to ensure that no mirror images are accidentally folded.

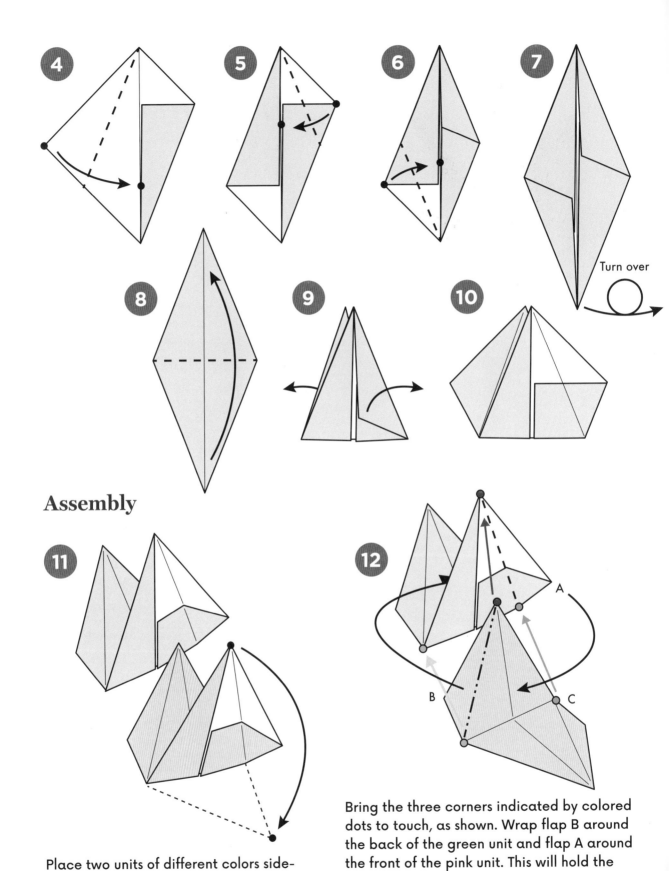

Turn over

Assembly

Place two units of different colors side-by-side, as shown. Lower the front corner.

Bring the three corners indicated by colored dots to touch, as shown. Wrap flap B around the back of the green unit and flap A around the front of the pink unit. This will hold the units together quite securely.

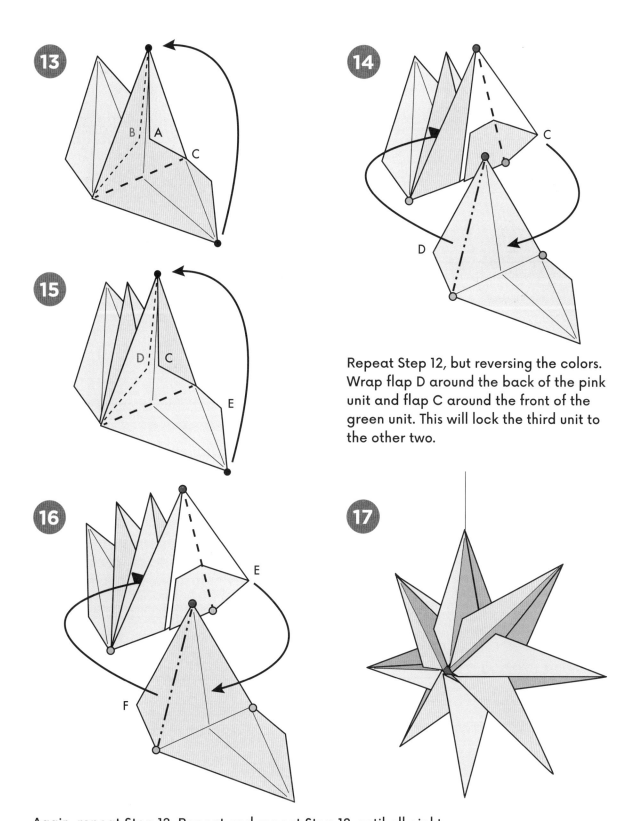

Repeat Step 12, but reversing the colors. Wrap flap D around the back of the pink unit and flap C around the front of the green unit. This will lock the third unit to the other two.

Again, repeat Step 12. Repeat and repeat Step 12, until all eight units have been locked in a line. To complete the Star, interlock the eighth unit with the first, completing a circle of linkages. Adjust the eight blades so that the Star is symmetrical.

Funny Faces

We can see faces in clouds, in burnt toast ... in fact, in anything! It's a phenomenon known as *pareidolia*. In origami, how much can we simplify a face so that we can still recognize it, albeit somewhat simplified and abstracted? Here are a few examples using just two or three folds. Drawing on the faces is optional, but it helps to give them some character. Can you create your own examples as simple as these? Fold as closely as you can to the locations indicated by the dots.

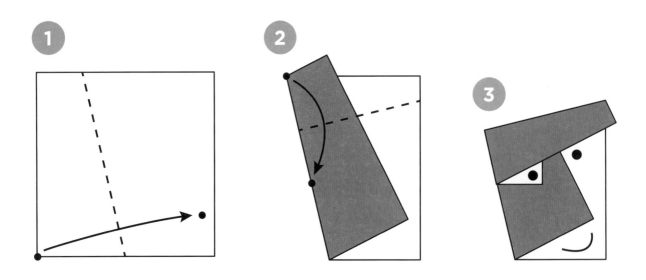

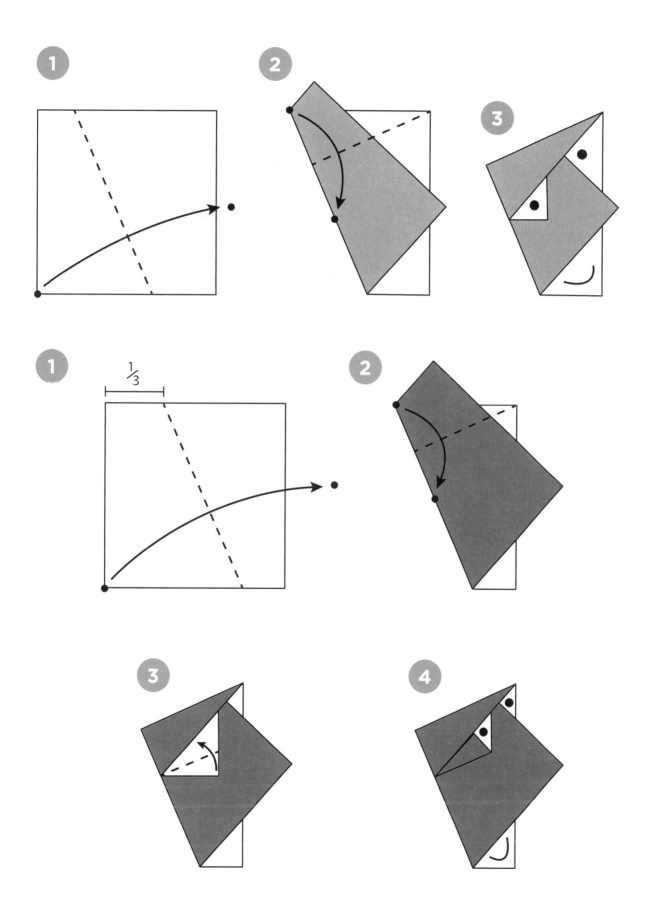

Feeding Bird

Birds are perhaps the most popular subject in origami, especially when the folding is simple. It's easy to see why: they are essentially a central blob with a pointy beak at one end and a pointy tail at the other. This generic shape is relatively easy to make in folded paper. The version shown here enables the bird to both stand and appear to feed.

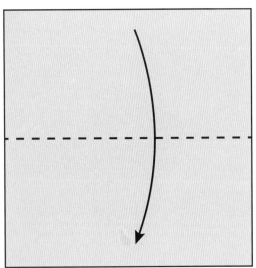

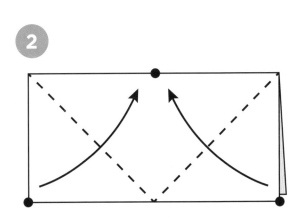

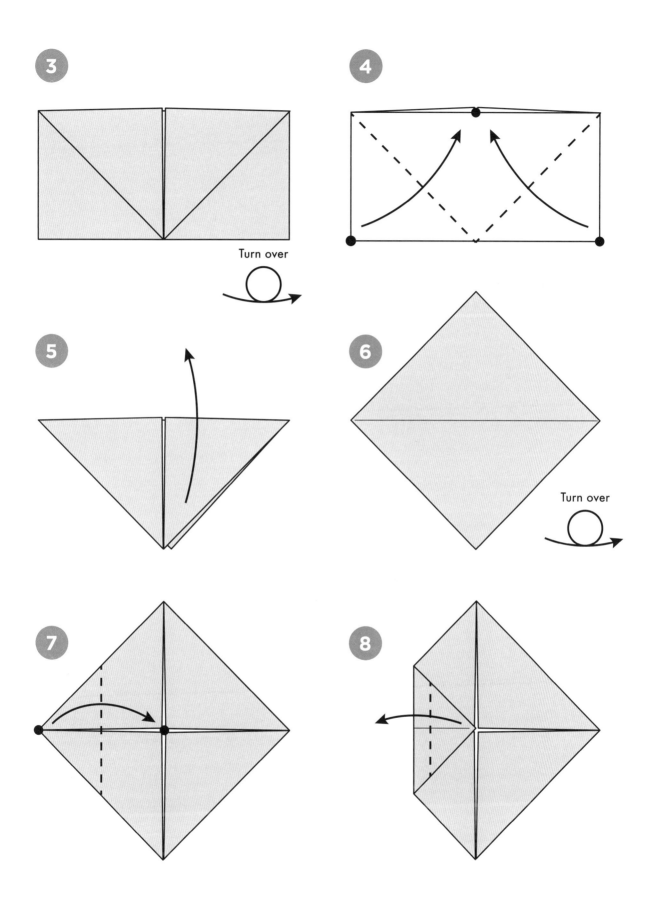

Turn over

Turn over

9

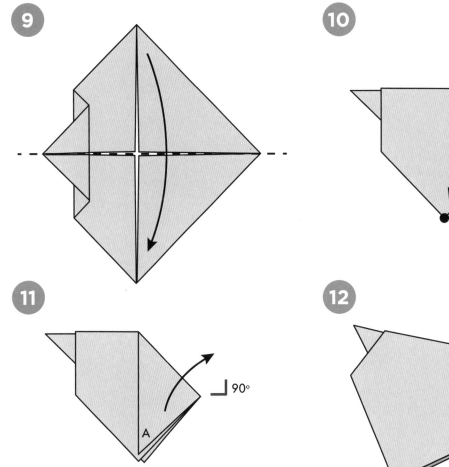

10

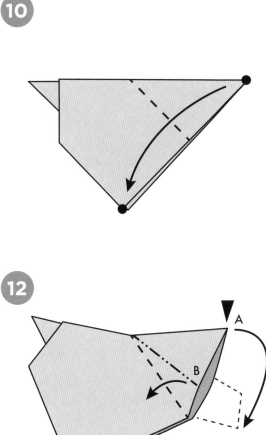

11

12

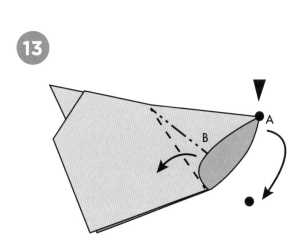

Lift up corner A so that it stands upright.

Apply pressure on A, opening the pocket beneath it.

13

14

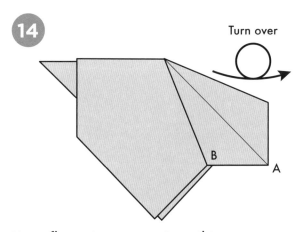

Keep flattening corner A, until it creates the kite shape shown here. Turn over.

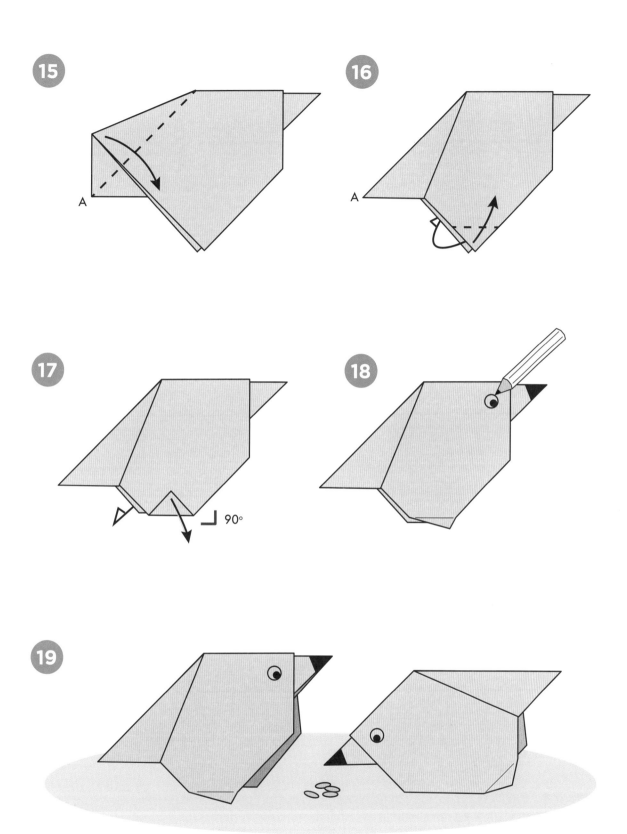

Shadow Dog

This design was commissioned by a paper museum in Denmark for an origami exhibition. The folding method was presented as a looping animation. Visitors to the museum could fold along and make their own model to take with them. Show the plain back to a friend and ask them what it is. They won't know the answer until you hold it up to the light and the silhouette of the dog is revealed!

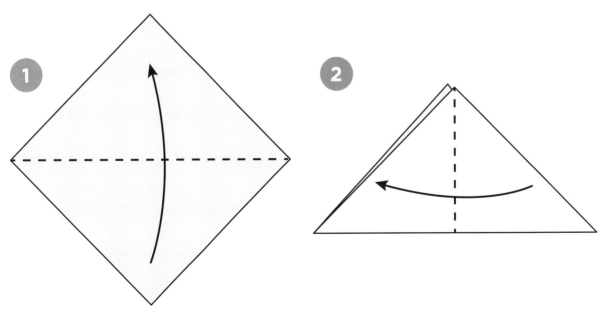

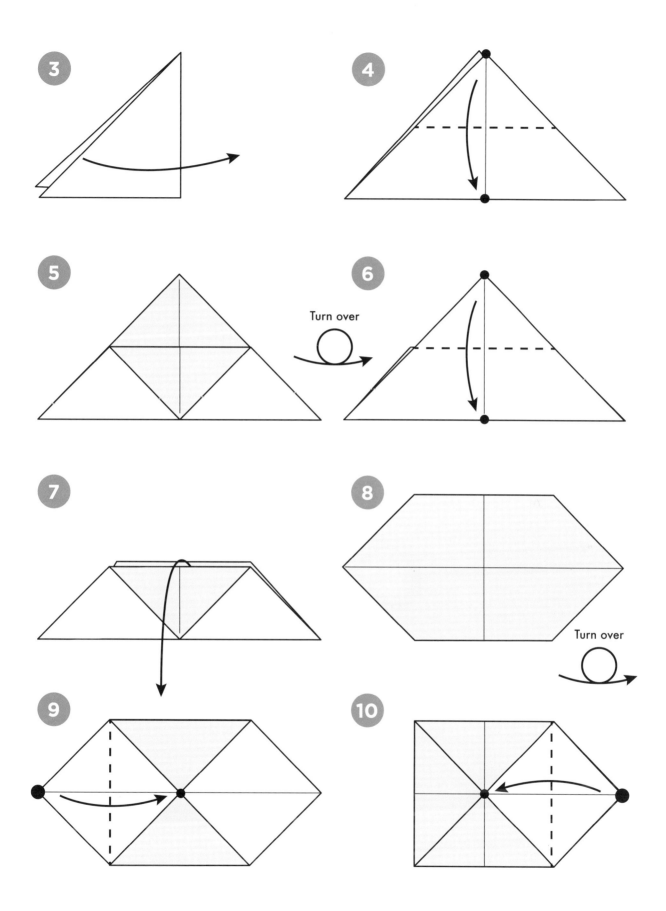

Turn over

Turn over

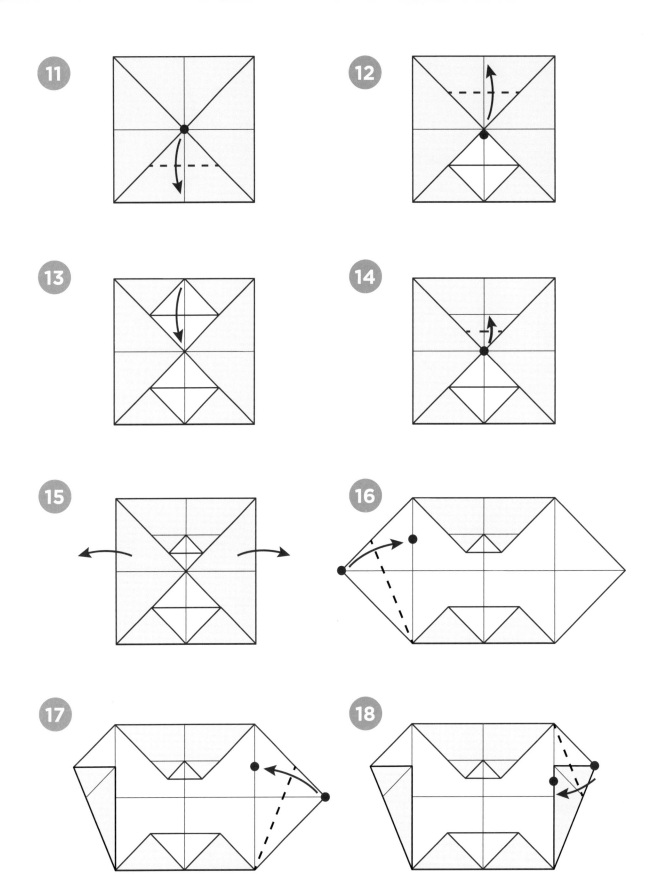

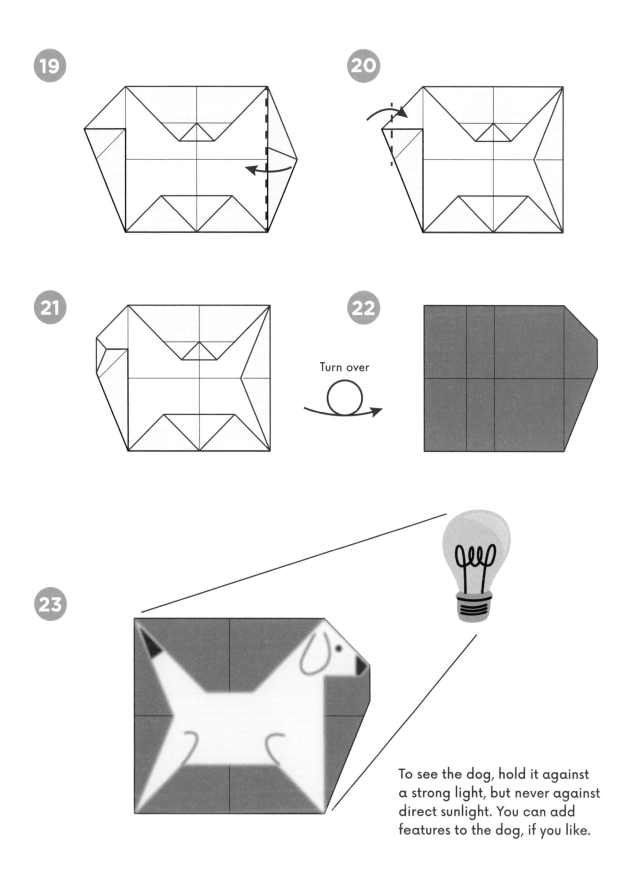

Turn over

To see the dog, hold it against a strong light, but never against direct sunlight. You can add features to the dog, if you like.

Butterfly

Butterflies abound in origami. Indeed, whole books have been written about them. I have created many myself, of varying levels of difficulty. This is one of my favorites, perhaps because of the unusual color-change technique that differentiates the body from the wings and because of the clean lines of the silhouette.

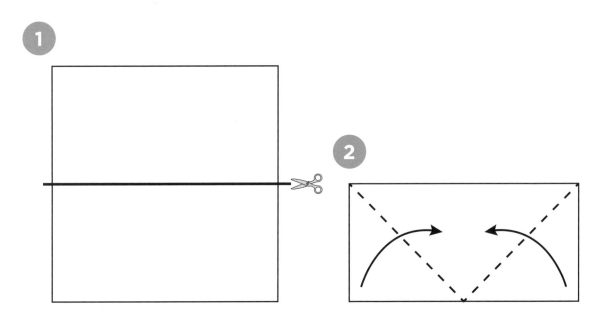

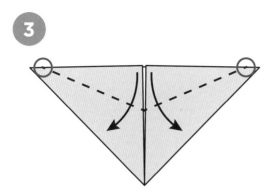

3

Note that at the top edges, neither fold originates at a corner. Look at Step 4 to see the result.

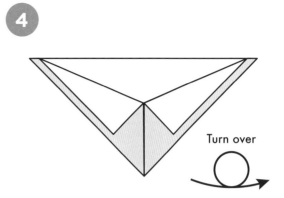

4

Turn over

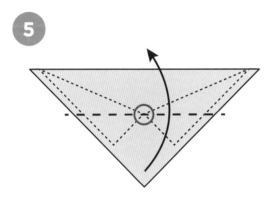

5

6

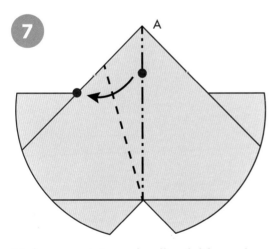

7

A

Make mountain and valley folds as shown, so that corner A folds in half and moves to the left.

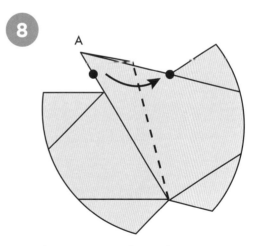

8

A

Swivel A across to the right.

9

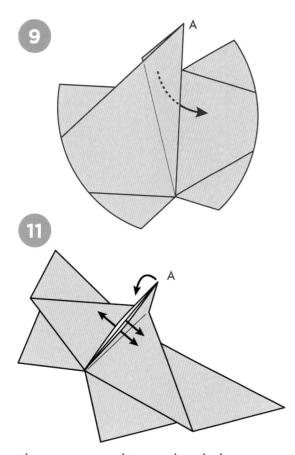

10

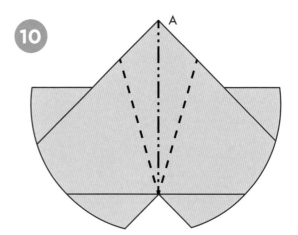

Make all three folds at the same time, so that corner A stands in the air.

11

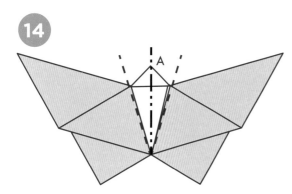

There are two white pockets below A, one to the left of the center fold and one to the right. Open one of them to expose white paper and squash A flat on top.

12

13

14

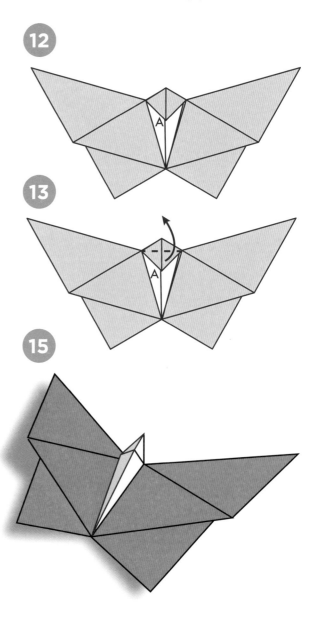

Refold the three creases from Step 10 so that the white body stands vertically.

15

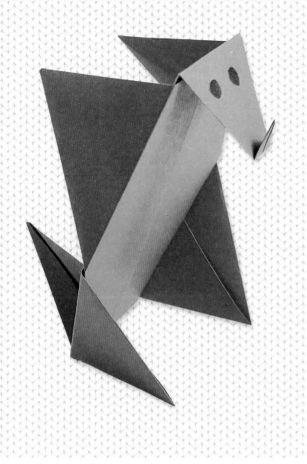

Foldzilla

This cute monster just appeared in my fingers one day when I was trying to create something very different. That's the way it goes sometimes: you're trying hard to achieve one thing ... and then another possibility comes along that turns out to be more interesting than what you were trying to make in the first place! Perhaps that's why many of my designs look like one-offs, not designed from a preconceived idea. Use two squares of the same color.

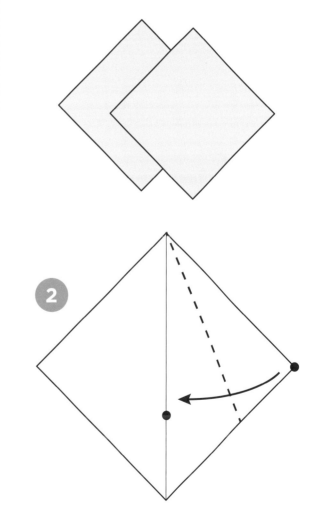

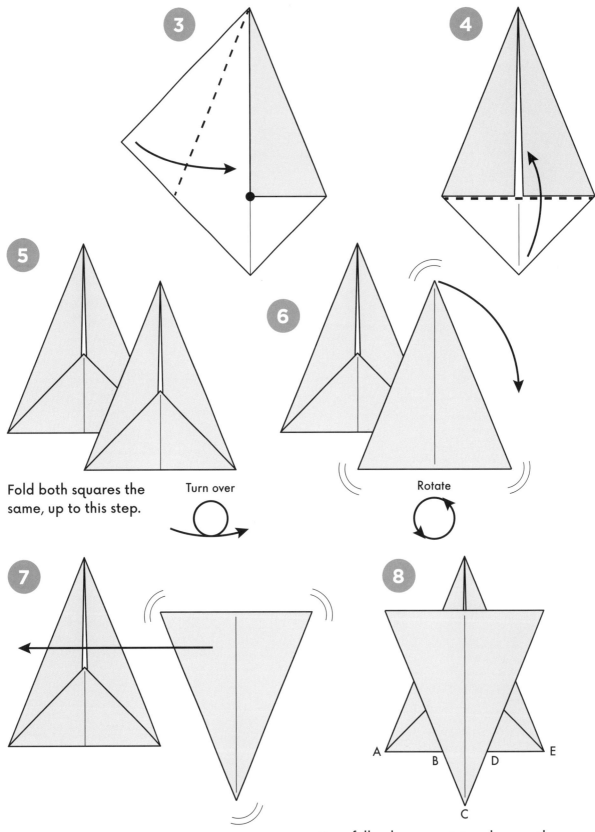

3

4

5

Fold both squares the same, up to this step.

Turn over

6

Rotate

7

8

A   B   D   E

C

Carefully place one triangle over the other, so that AB = BC = CD = DE.

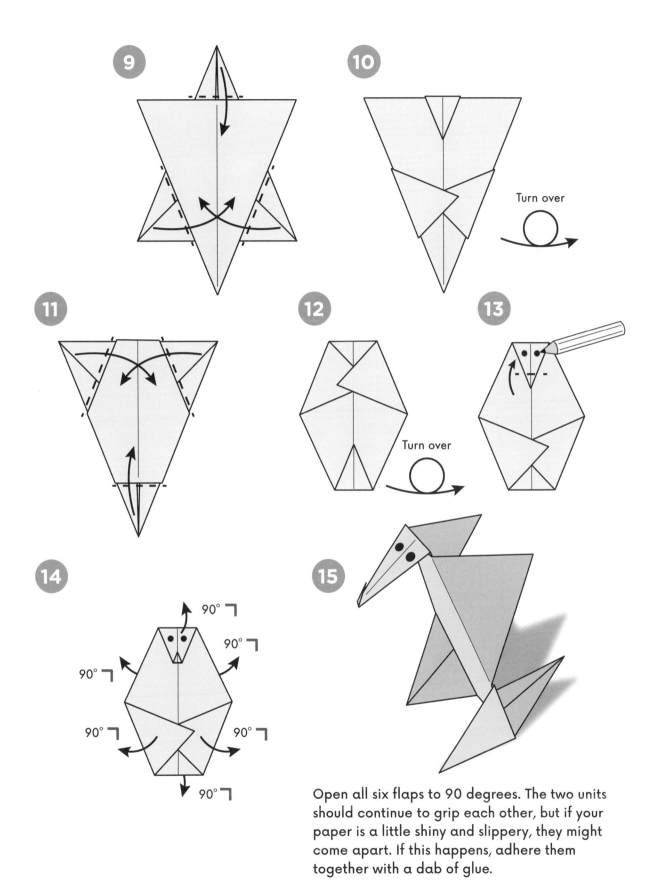

9

10

Turn over

11

12

Turn over

13

14

90°

90°

90°

90°

90°

90°

90°

15

Open all six flaps to 90 degrees. The two units should continue to grip each other, but if your paper is a little shiny and slippery, they might come apart. If this happens, adhere them together with a dab of glue.

Penguin

Almost every origami designer has their own penguin or three! This is likely because the basic form of the bird is easy to create, it has an interesting and necessary color-change, and penguins are just cute and people like to fold them! Its distinctive black and white coloring means that liberties can be taken with the form of the design and it will still be recognizable.

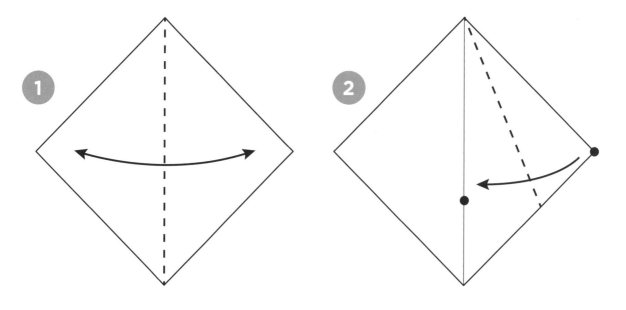

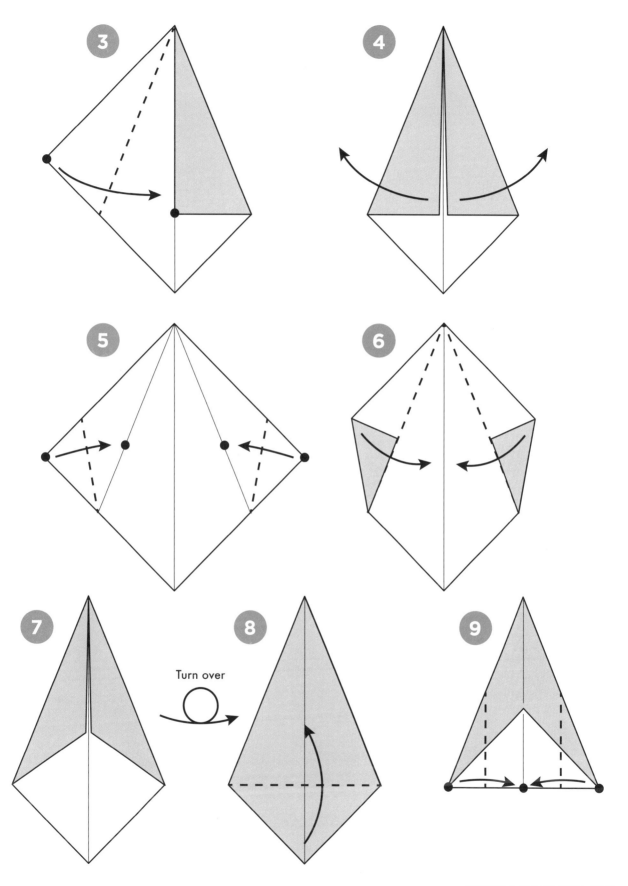

Turn over

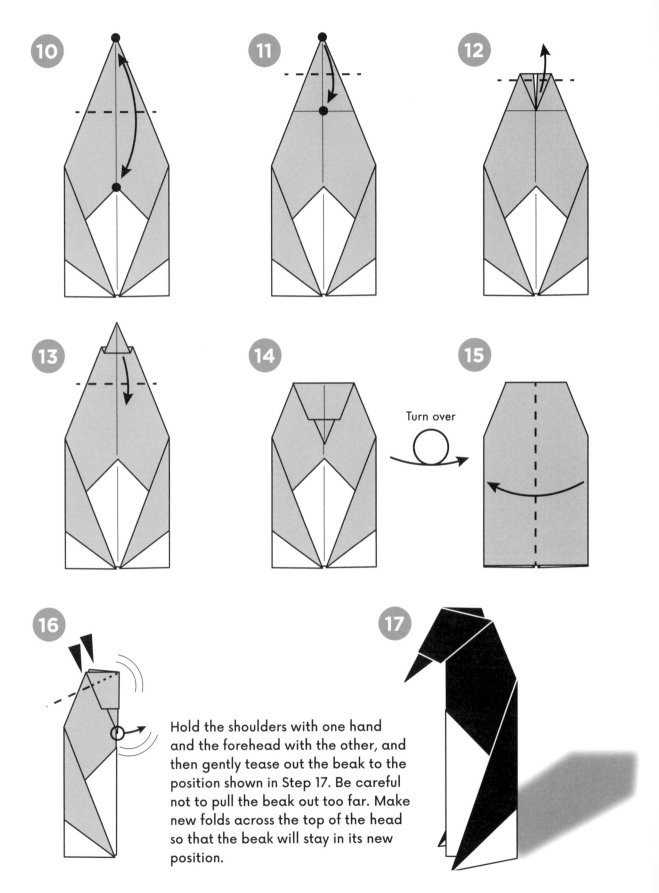

Hold the shoulders with one hand and the forehead with the other, and then gently tease out the beak to the position shown in Step 17. Be careful not to pull the beak out too far. Make new folds across the top of the head so that the beak will stay in its new position.

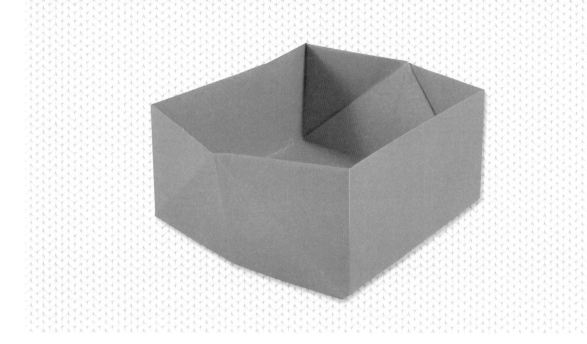

Instant Box No. 1

This is the first of two related boxes that look the same when collapsed. They also open the same way and look the same once open. But they are folded in very different ways! This one uses a grid that's aligned with the sides of the paper. Fold this box first, and then fold the second box (page 82) to compare and contrast the two. Origami is not just about the "what"—it's also about the "how."

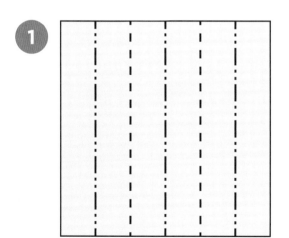

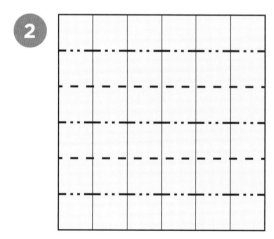

In Steps 1 and 2, the paper is divided into a 6 × 6 grid. For simplicity, you can make an 8 × 8 grid, and then trim away quarter-width strips from two adjacent edges. To make an 8 × 8 grid, make half and quarter folds in each direction, and then fold each quarter-width strip in half. Note above which folds are mountains and valleys.

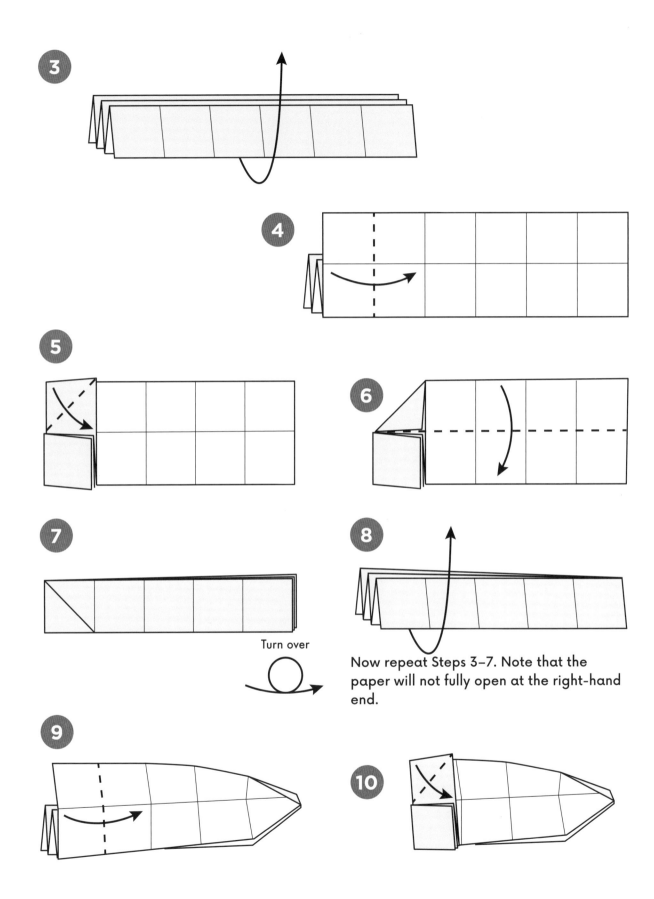

3

4

5

6

7

Turn over

8

Now repeat Steps 3–7. Note that the paper will not fully open at the right-hand end.

9

10

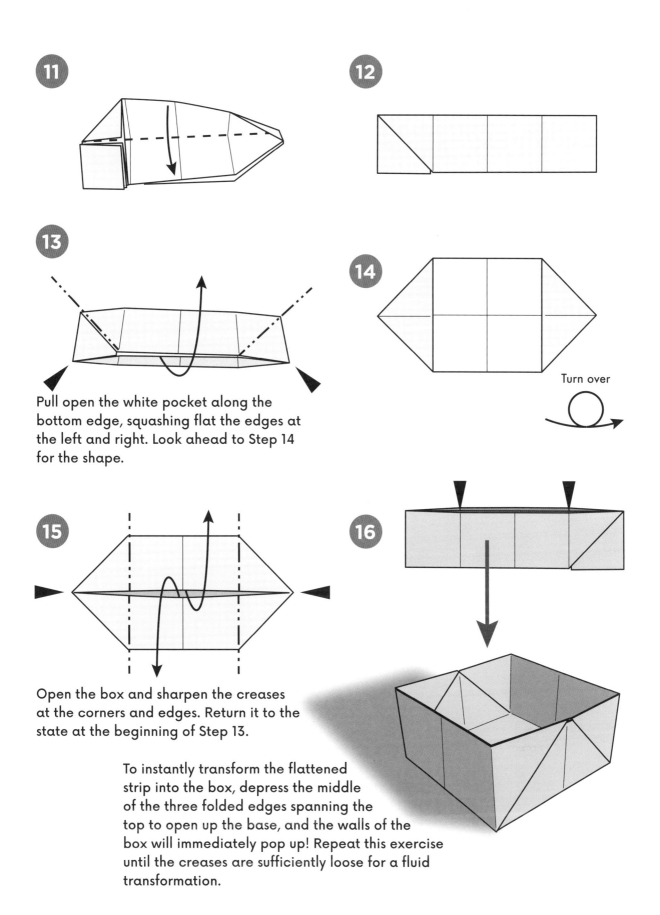

11

12

13

Pull open the white pocket along the bottom edge, squashing flat the edges at the left and right. Look ahead to Step 14 for the shape.

14

Turn over

15

Open the box and sharpen the creases at the corners and edges. Return it to the state at the beginning of Step 13.

To instantly transform the flattened strip into the box, depress the middle of the three folded edges spanning the top to open up the base, and the walls of the box will immediately pop up! Repeat this exercise until the creases are sufficiently loose for a fluid transformation.

16

Instant Box No. 2

In contrast to Instant Box No. 1, this box is made using a grid running at 45 degrees to the edges of the paper. Although the two boxes look alike and open and close in the same manner, the folding sequence here is more unorthodox, the layers are more evenly distributed, and there's a color change. Instantaneously transforming both boxes from flat, to 3-D and back is fun—and somewhat magical!

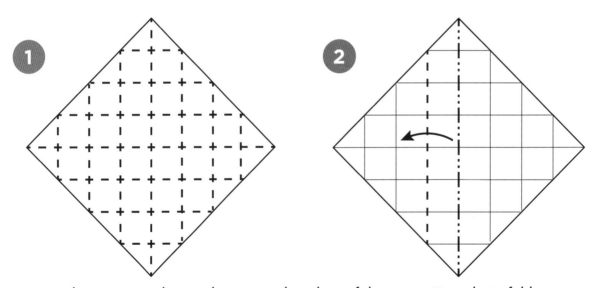

Begin with an 8 × 8 grid at 45 degrees to the edges of the paper. To make it, fold corner to corner in both directions, and then fold each corner to the center of the paper and unfold. Then, install creases between each corner and its closest interior crease intersection. Finally, install creases between the perpendicular central creases and their closest parallel creases.

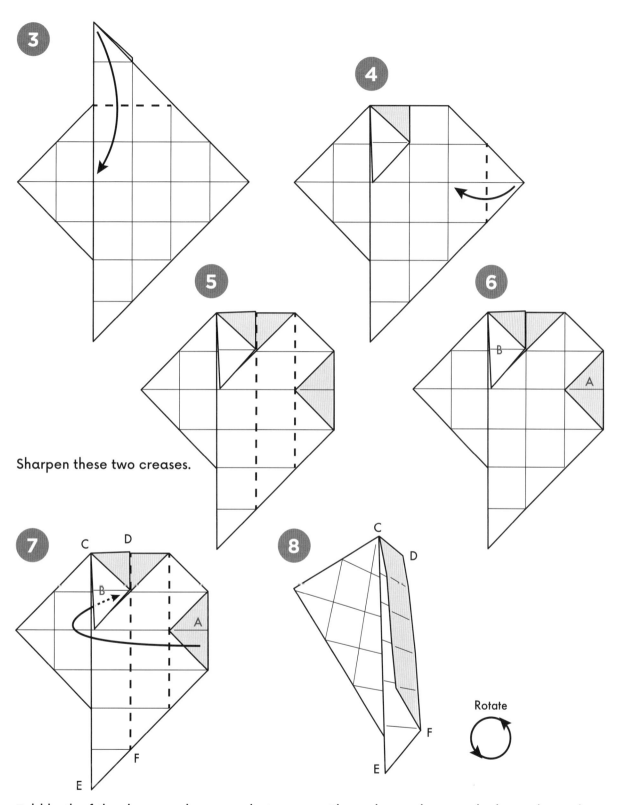

Sharpen these two creases.

Fold both of the sharpened creases, being careful to tuck flap A immediately behind flap B, NOT behind edge CE.

This is the result. Note the long white edge, CE. Rotate the paper 180 degrees.

Rotate

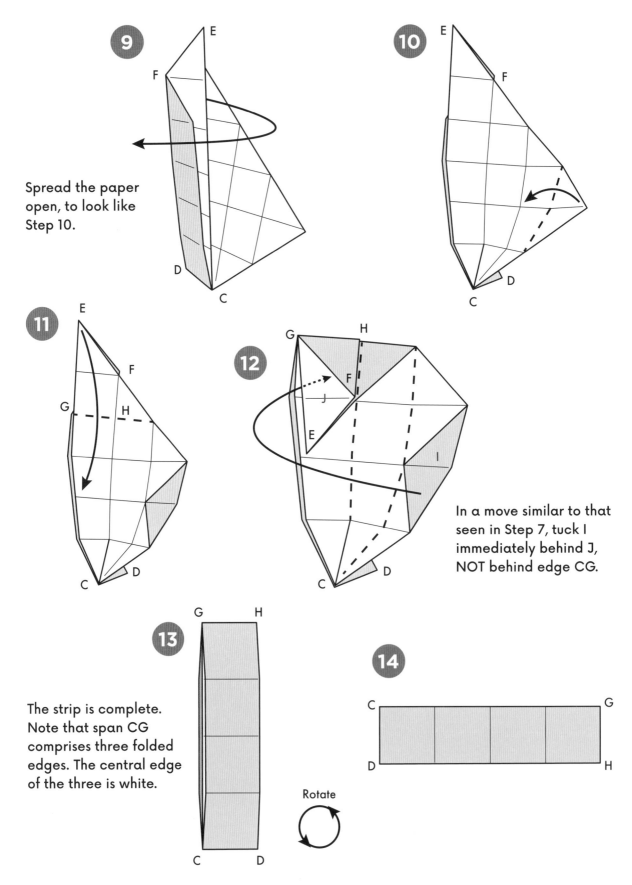

9

Spread the paper open, to look like Step 10.

10

11

12

In a move similar to that seen in Step 7, tuck I immediately behind J, NOT behind edge CG.

13

The strip is complete. Note that span CG comprises three folded edges. The central edge of the three is white.

Rotate

14

15

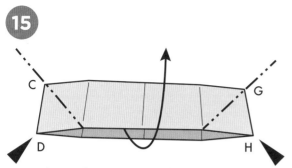

Note that edge CG is at the top. Open and spread edge DH, creating triangles at the left and right.

16

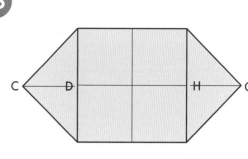

Turn over

17

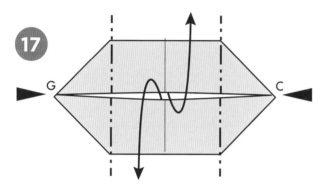

Open the box and sharpen the creases at the corners and edges.

18

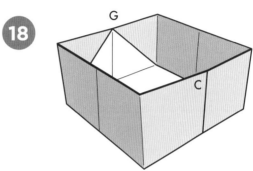

Return the box to the state at the beginning of Step 15.

19

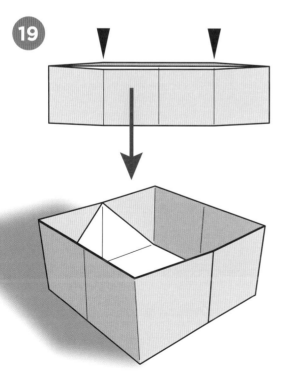

To instantly transform the flattened strip into the box, depress the middle of the three folded edges spanning the top to open up the base, and the walls of the box will immediately pop up (just as with Instant Box No. 1)! Repeat this exercise until the creases are sufficiently loose for a fluid transformation.

Kevin's Boat

This lovely design was created by my nephew, Kevin Golan-Aharonov, when he was only six! I like the simple and direct folding sequence, the way it stands up, and the color change option presented by the sails. Try experimenting with the placement of the folds in Steps 1–4 to create boats with different proportions. You can even change the proportions of the paper for even more options.

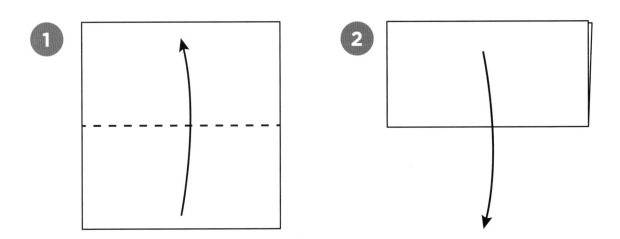

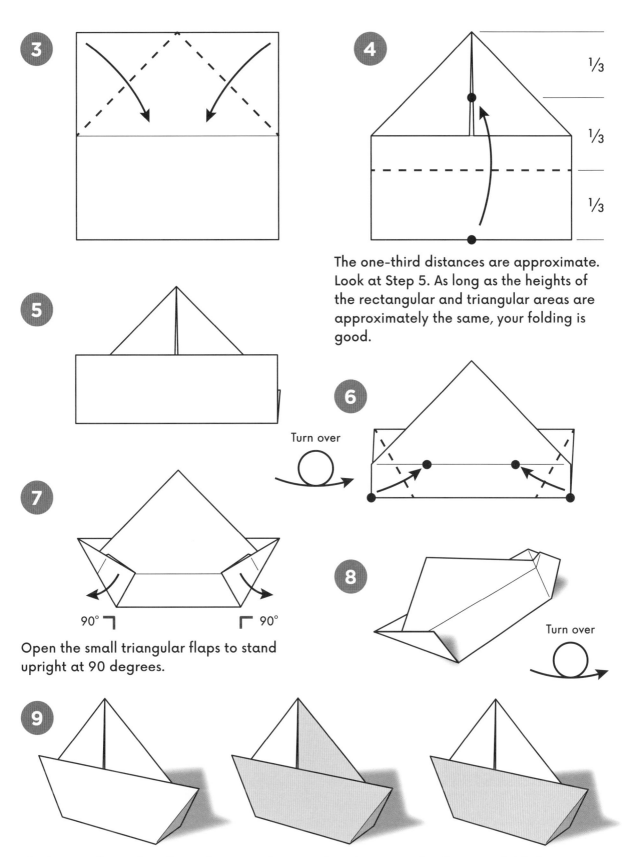

The one-third distances are approximate. Look at Step 5. As long as the heights of the rectangular and triangular areas are approximately the same, your folding is good.

Open the small triangular flaps to stand upright at 90 degrees.

The boat will stand well. By changing the way that the triangular flaps are folded in Step 3, the sails can have different color combinations.

Sneaker

This is one of the few designs in the book that was designed intentionally, after I saw a photograph of a side-on sneaker and realized, to my surprise, that the shape was unexpectedly chunky—and thus, relatively easy to fold. All I needed to do was move the main folds around until, eventually, the basic design became simple to make. The bonus is that you can decorate them as you wish!

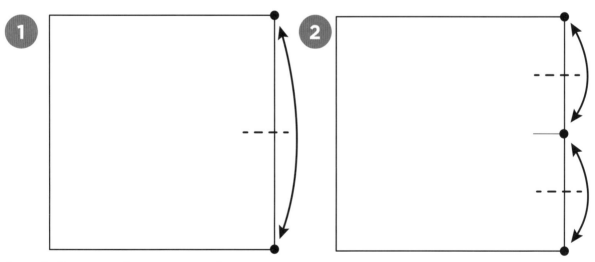

Steps 1–5 consist of a sequence of precise pinched landmarks. Be sure to follow the directions carefully.

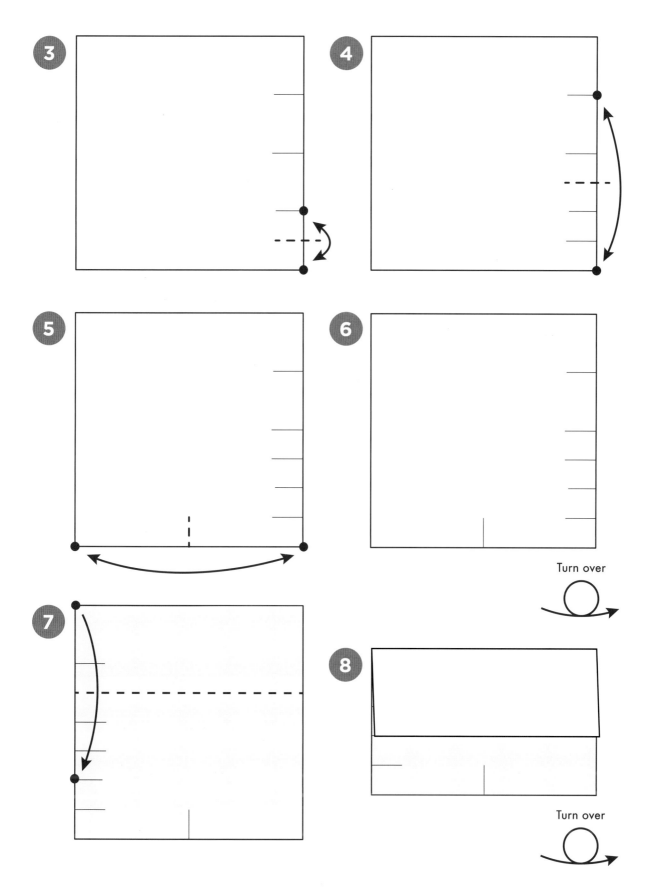

Turn over

Turn over

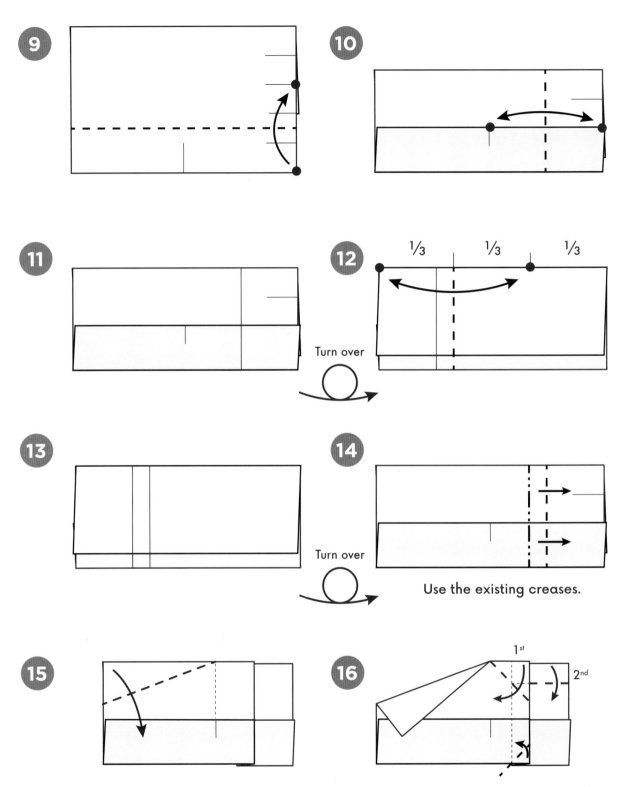

9

10

11

12 ⅓ ⅓ ⅓

Turn over

13

14

Turn over

Use the existing creases.

15

Note the precise location of the fold.

16 1st 2nd

At the top edge, make the two folds in the order shown. Note that a small triangular pocket will form where the two folds meet.

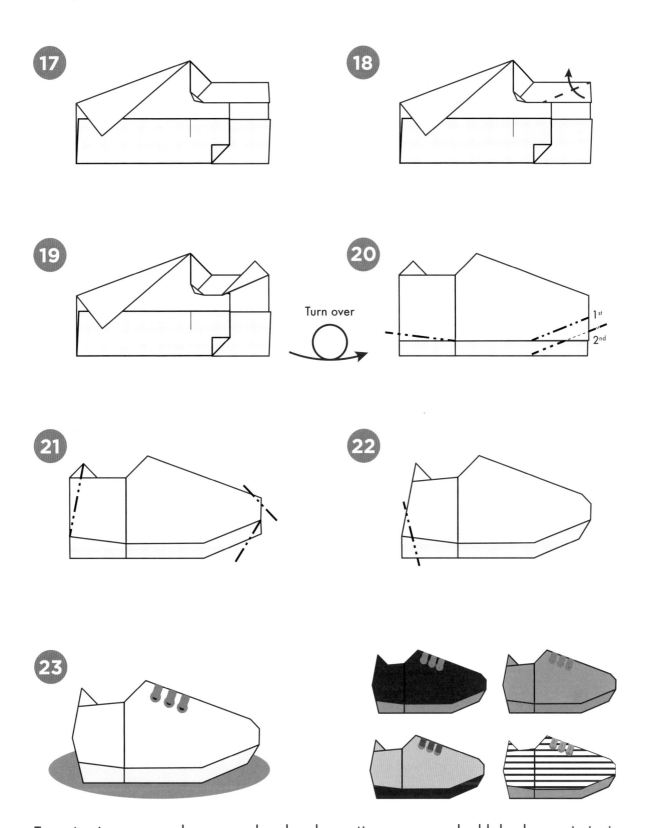

Turn over

To customize your sneakers, use colored or decorative papers, and add shoelace or insignia details.

Arrow

This design can be created from any oblong rectangle. Longer rectangles will yield longer arrow shafts. The instructions here are for a monochrome arrow, which requires paper with the same color on the front and back, but you can make a color-change design if you use paper with different colors on the two surfaces. If you want a challenge, try making an arrow with a head at both ends, with a narrower shaft, or with a pointier head.

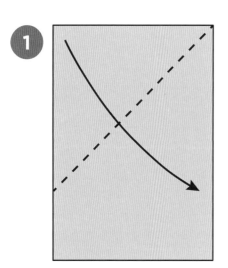

Use 8½ × 11-inch paper, A4, or any other oblong rectangle.

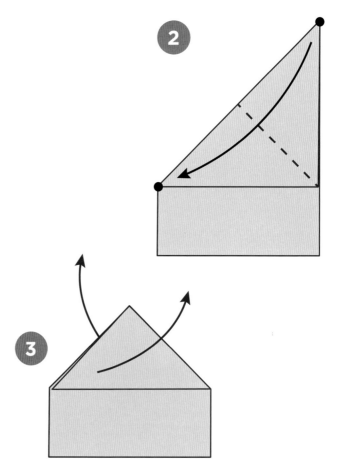

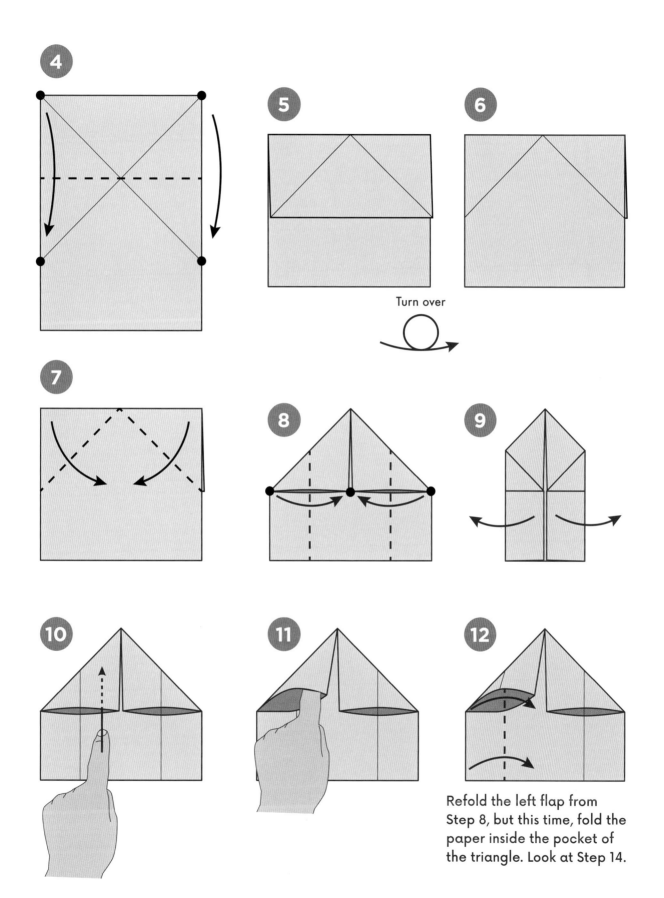

Turn over

Refold the left flap from Step 8, but this time, fold the paper inside the pocket of the triangle. Look at Step 14.

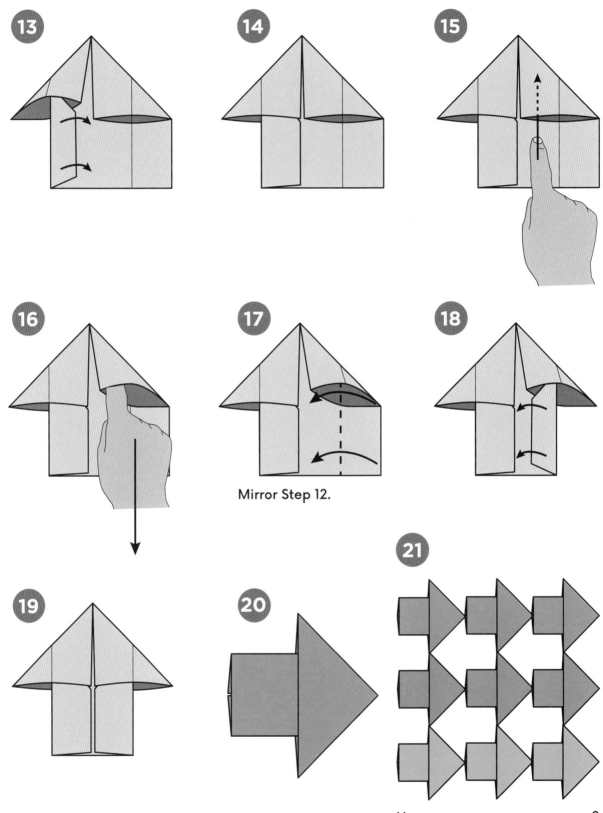

13

14

15

16

17

Mirror Step 12.

18

19

20

21

How many arrows can you see?

Lark Box

My dear origami friends Larry Hart and Mark Bolitho didn't make it through the pandemic. Both were top creators. Larry became famous as a teenager some decades ago, having created a classic Instant Cube. Mark later made a variation of this design, featuring a fancy opening to a cubic pot. My design is mid-way between the two. For this reason, I have combined LA-rry and Ma-RK's names and called it the Lark Box. If available, use paper larger than 6 × 6 in (15 × 15 cm).

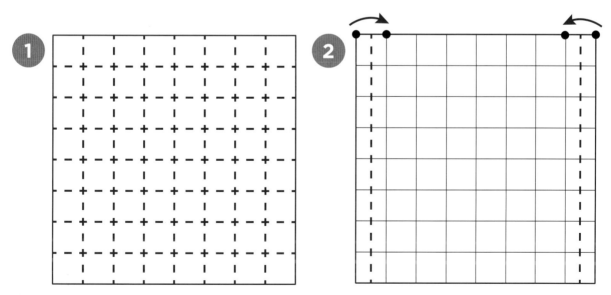

The box is divided into a 8 × 8 grid. To do this, fold and unfold in half, horizontally and vertically. Then, fold and unfold the quarters, horizontally and vertically. Finally, add creases between all of the existing creases.

3

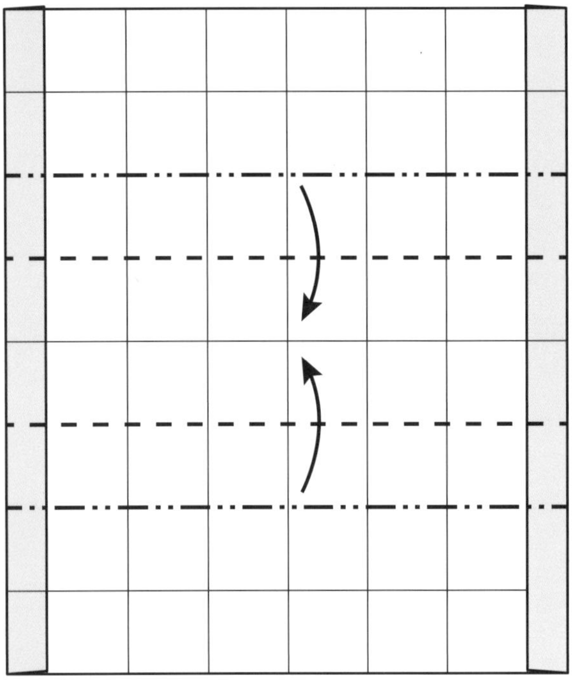

4

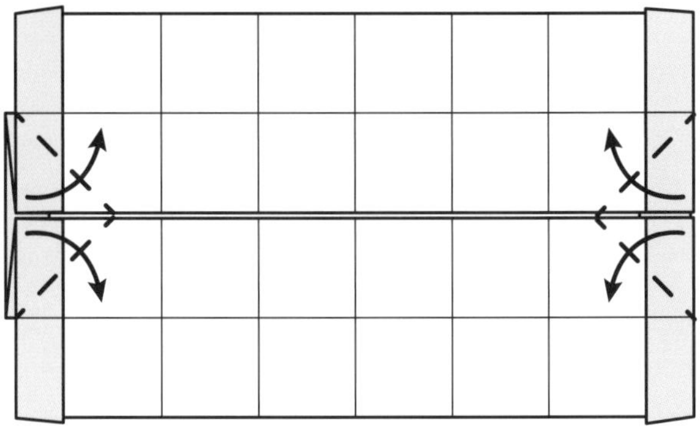

5

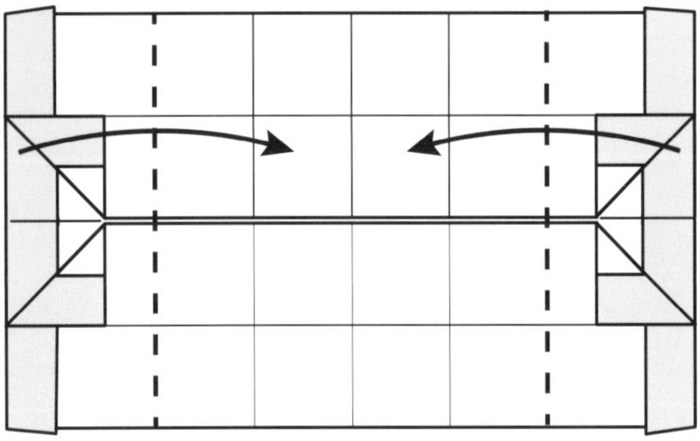

6

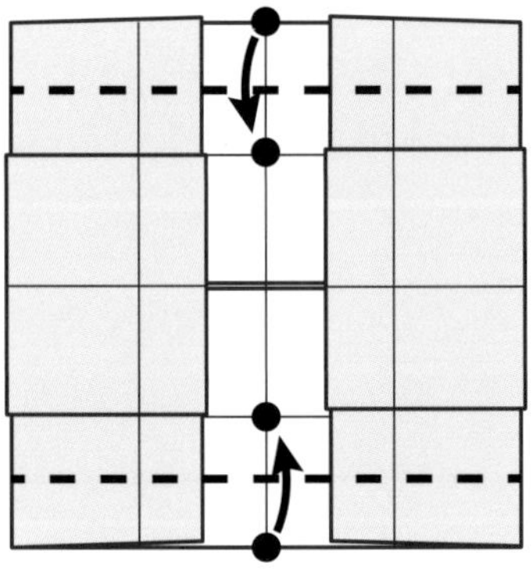

7

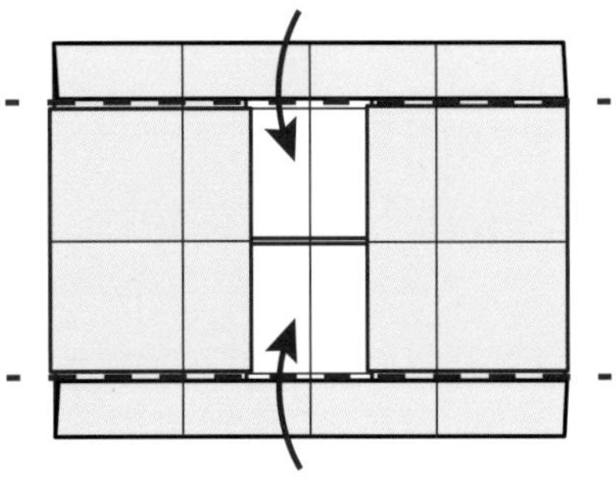

8

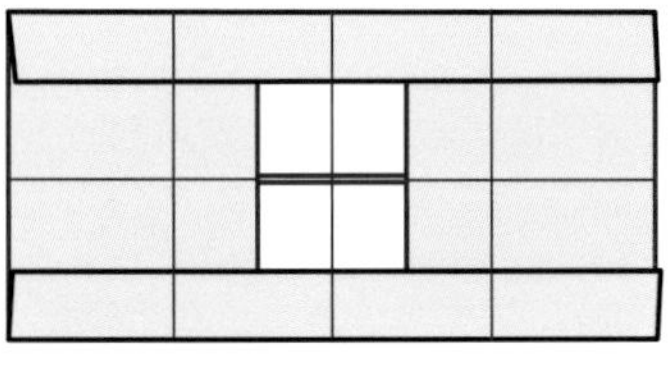

9

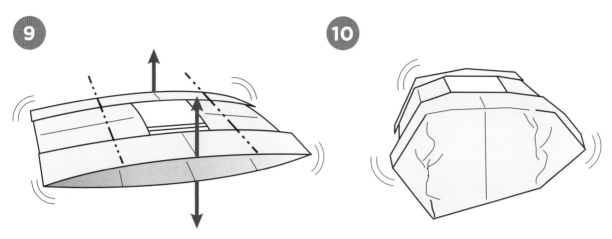

Begin to create volume inside the box. Note the mountain folds.

10

The vertical edges of the box can be messy and difficult to make neat and tidy. If this is happens to you, Step 11 has the solution.

11

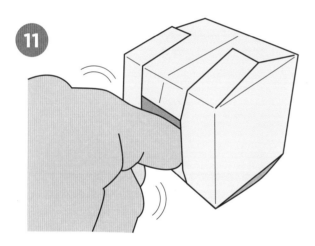

Lay the box on its side. Put a finger through the hole and begin to press the creases flat against the surface below the box. This method will enable you to tidy all the edges.

12

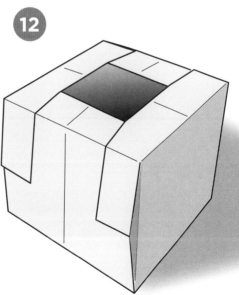

Tent

This design requires the precise geometry of A4 paper, so that the roof of the tent will fit exactly onto the triangular front and back sections. The instructions here are for a monochrome tent, which requires paper with the same color on the front and back. A4 yields a large model, but you can make a smaller rectangle by cutting an A4 sheet in half or even into quarters. Observe the mathematical relationship between the lengths of the edges that form the snugly-fitting components.

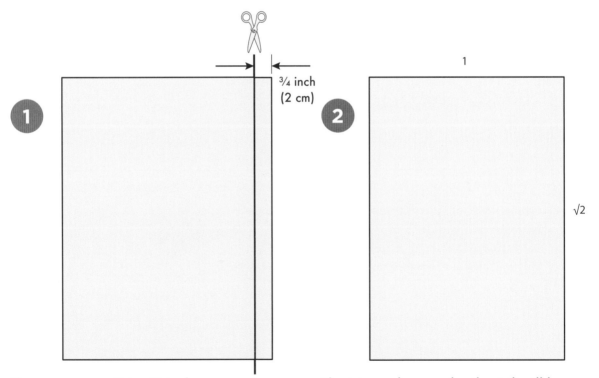

If you are using 8½ × 11-inch paper, trim ¾ inch (2 cm) from the long side to create a rectangle proportioned the same as A4 paper. If you are using A4, skip this step.

The trimmed paper (or the A4) will have this proportion.

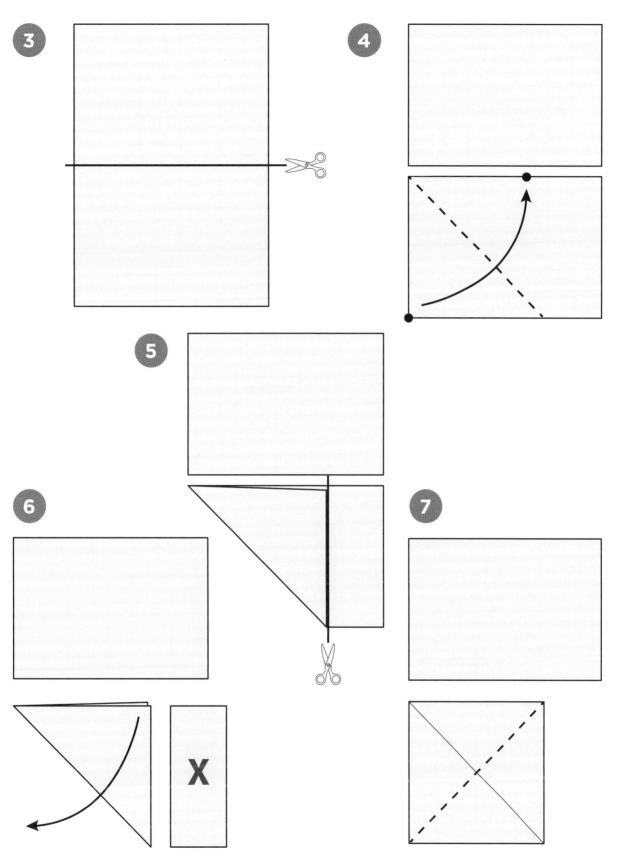

Discard the rectangle with the X.

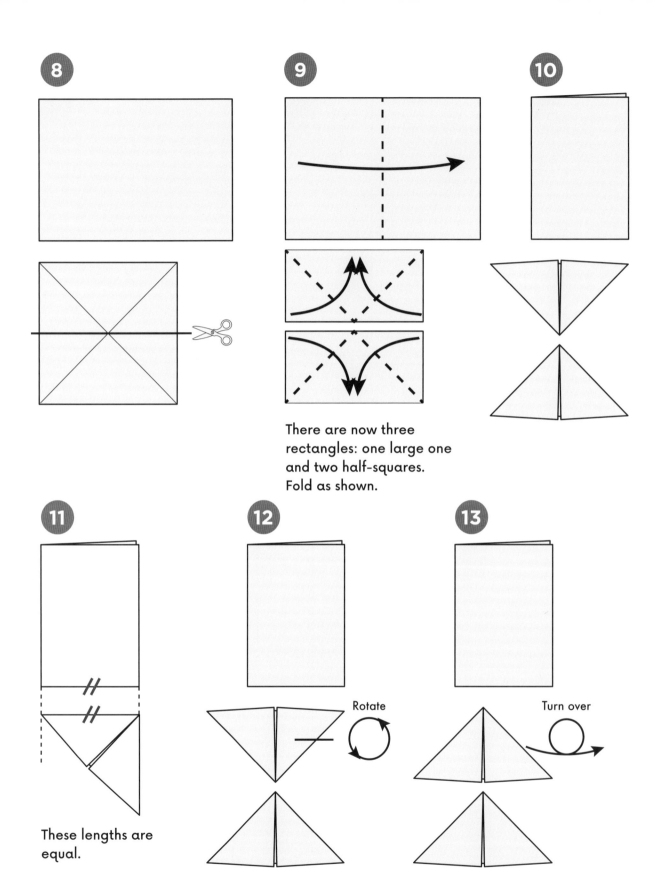

8

9

There are now three rectangles: one large one and two half-squares. Fold as shown.

10

11

These lengths are equal.

12

Rotate

13

Turn over

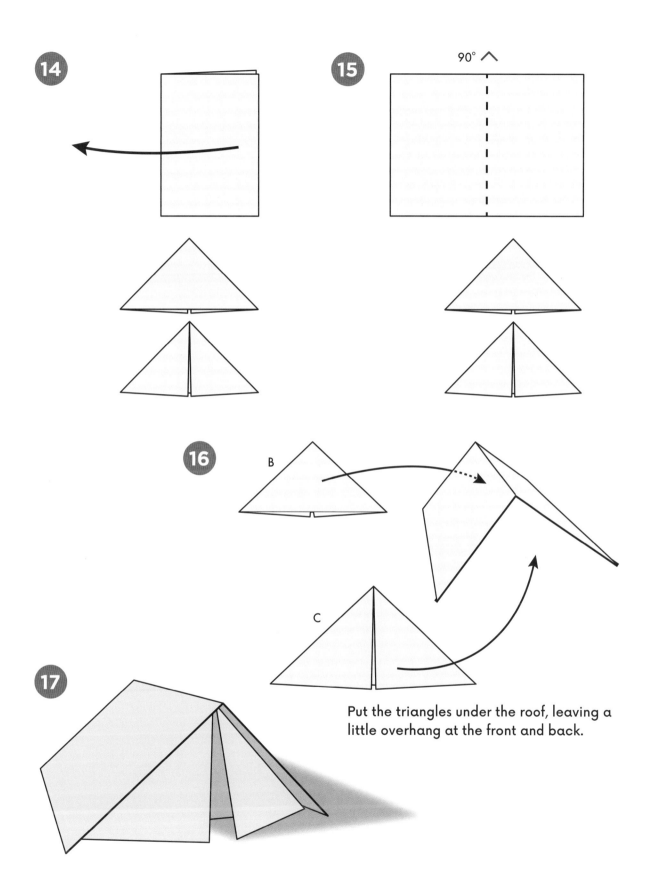

Put the triangles under the roof, leaving a little overhang at the front and back.

Vehicles

This suite of four designs all feature the same technique to fold the wheels, but they differ in the way the square of paper is initially divided. These variations create different options for vehicle shapes, ranging from a chunky cargo van to a sleeker sport sedan. The extreme economy with which the designs are made shows that sometimes, less is more (more or less!).

Sedan

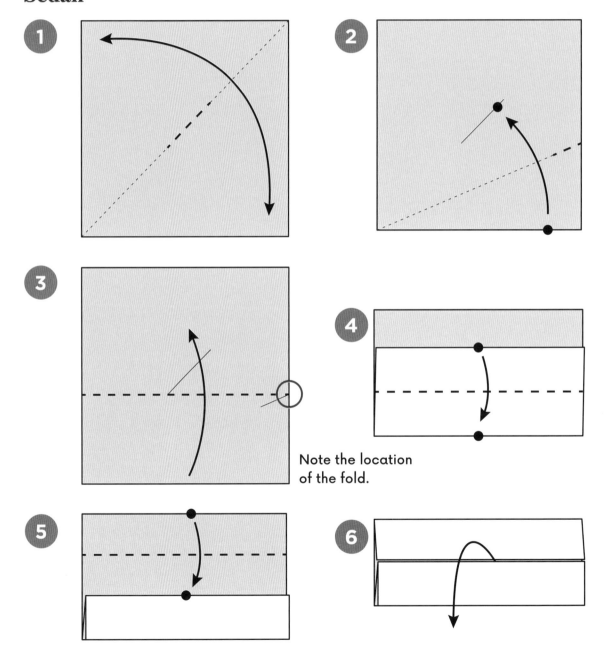

Note the location of the fold.

Sedan

Minivan

SUV

Cargo Van

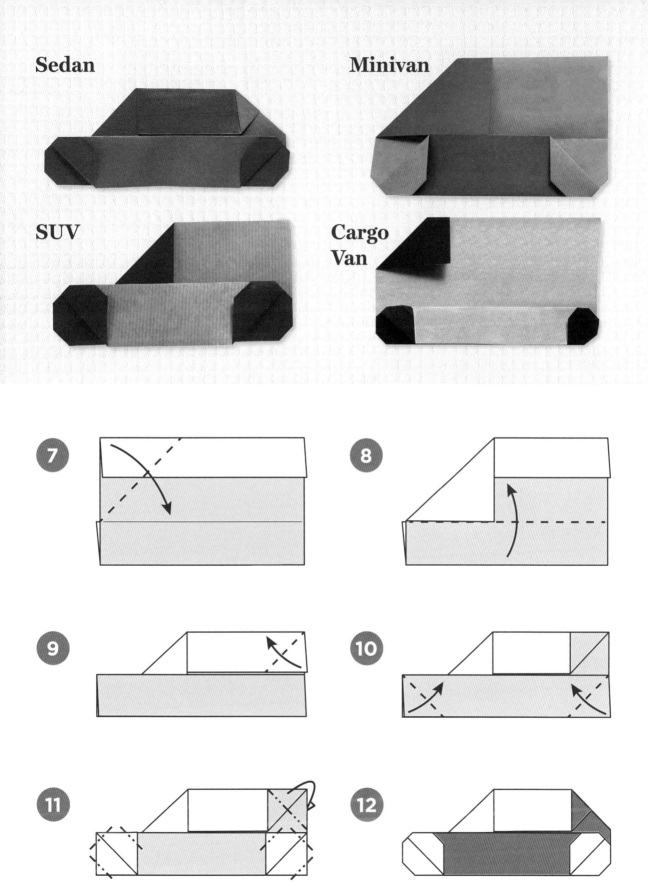

7

8

9

10

11

12

Minivan

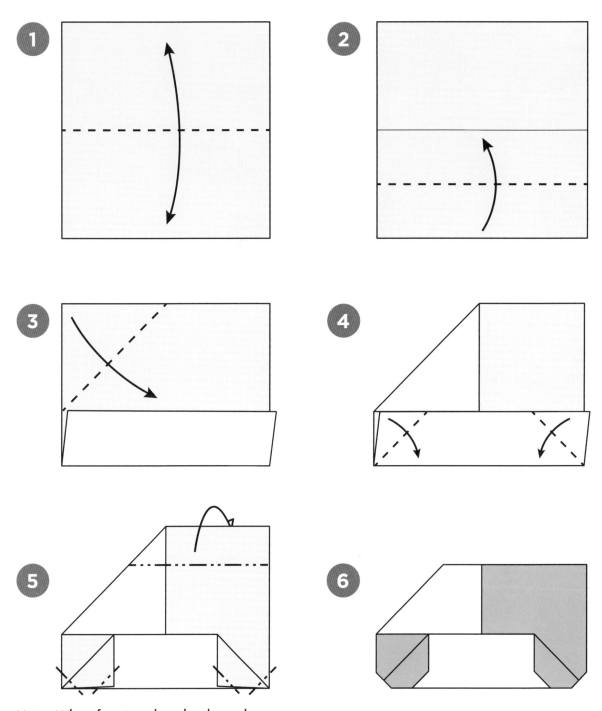

Note: When forming the wheels on the vehicles, take care when rounding them off—don't make them either too pointed or too square.

SUV

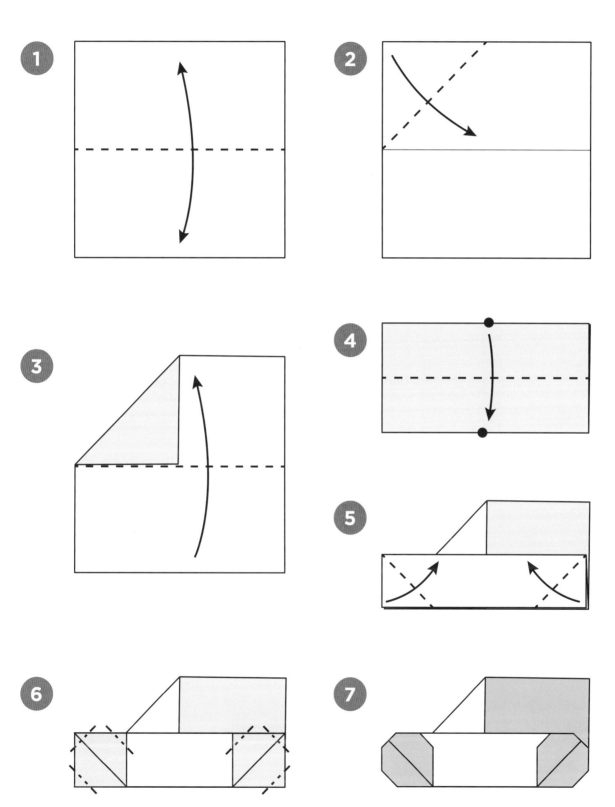

Cargo Van

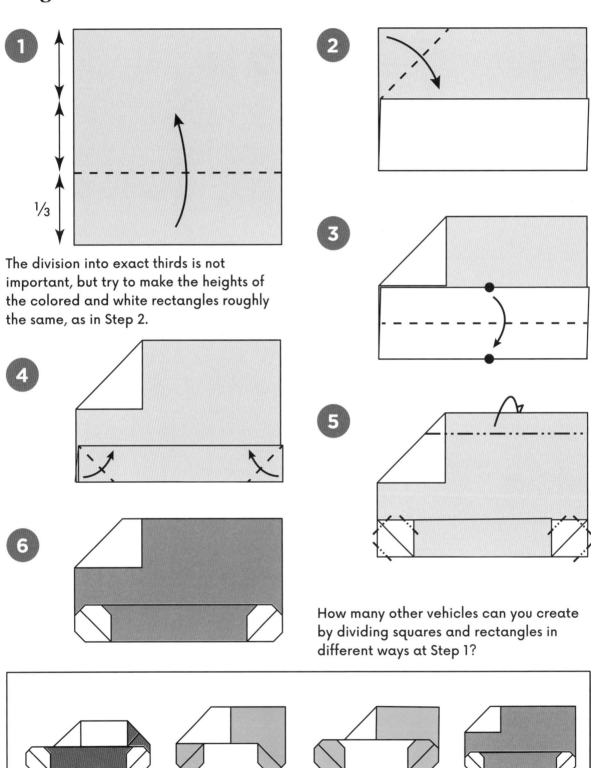

1 ⅓

The division into exact thirds is not important, but try to make the heights of the colored and white rectangles roughly the same, as in Step 2.

2

3

4

5

6

How many other vehicles can you create by dividing squares and rectangles in different ways at Step 1?

Pajarita Envelope

All of the designs in this book are new—except for this one, which has been sitting in a pile of scribbly drawings for perhaps forty years. I was reminded of it after seeing something similar online. The Pajarita ("Little Sparrow") is a traditional Spanish origami design of some antiquity, well-known in the country. The silhouette in the middle of Pajarita Envelope is the shape of the bird.

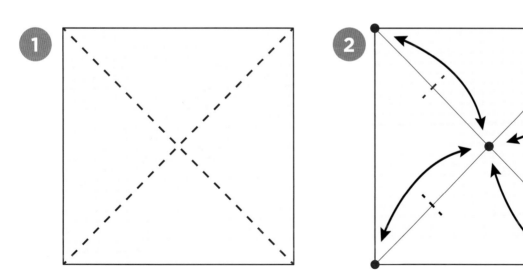

In Steps 1–6, complete the pre-creasing and the folding of the corners (they are not all folded the same!) very precisely.

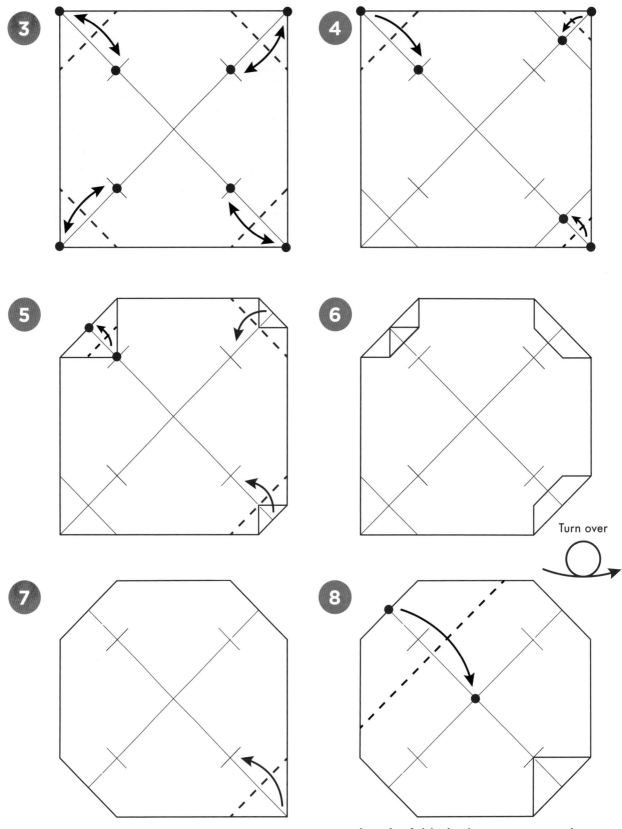

Turn over

Note that the folded edge is opposite the colored triangle.

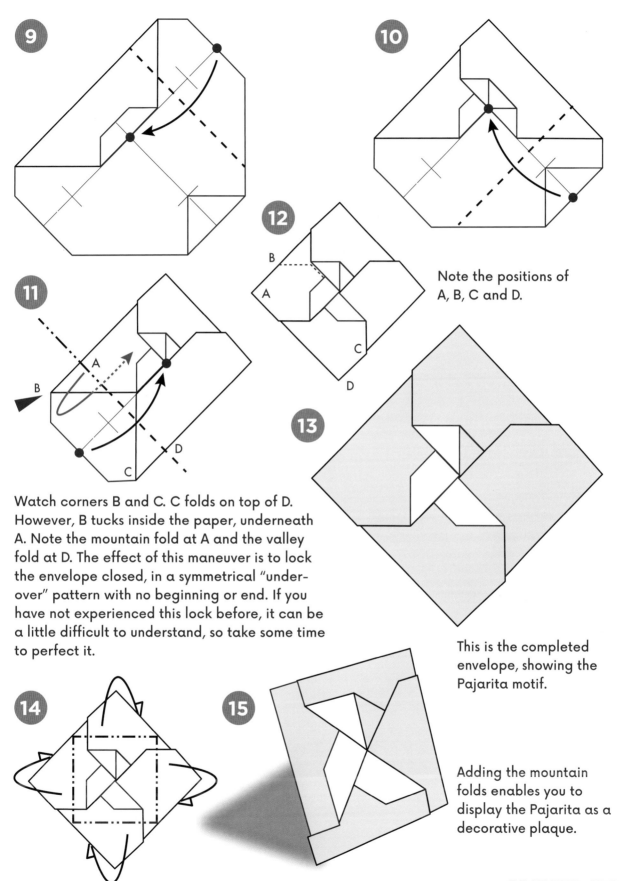

9

10

11

12

Note the positions of A, B, C and D.

Watch corners B and C. C folds on top of D. However, B tucks inside the paper, underneath A. Note the mountain fold at A and the valley fold at D. The effect of this maneuver is to lock the envelope closed, in a symmetrical "under-over" pattern with no beginning or end. If you have not experienced this lock before, it can be a little difficult to understand, so take some time to perfect it.

13

This is the completed envelope, showing the Pajarita motif.

14

15

Adding the mountain folds enables you to display the Pajarita as a decorative plaque.

Coffee Mug

The main technical challenge here was to create a faceted and locked cylinder with an extra "ear" for the handle, so the folding needs to be precise for the final result to look symmetrical. It can be made from any oblong rectangle, but either an A4 or 8½ × 11-inch sheet will yield a mug-like result. There is a variation with a closed base, but that model has waaay too many steps to fit into this book!

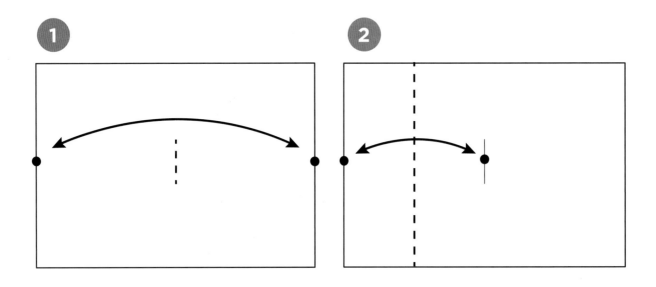

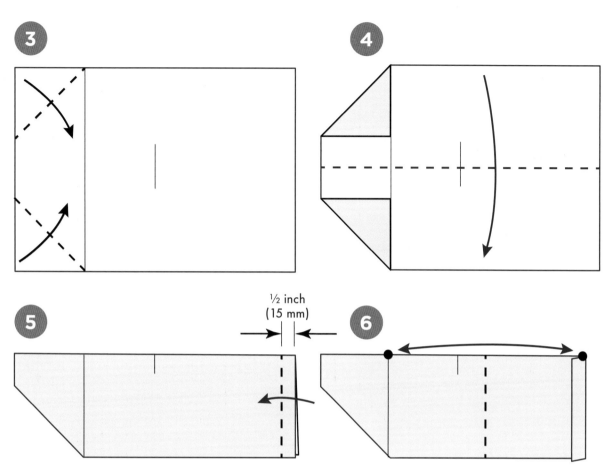

½ inch (15 mm)

Fold over the right-hand edge, as shown. Be sure to keep it folded until it is unfolded at Step 9.

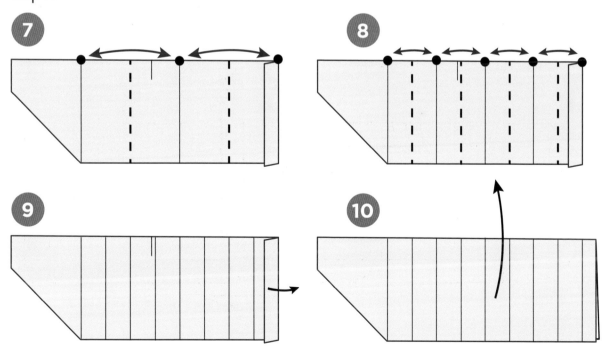

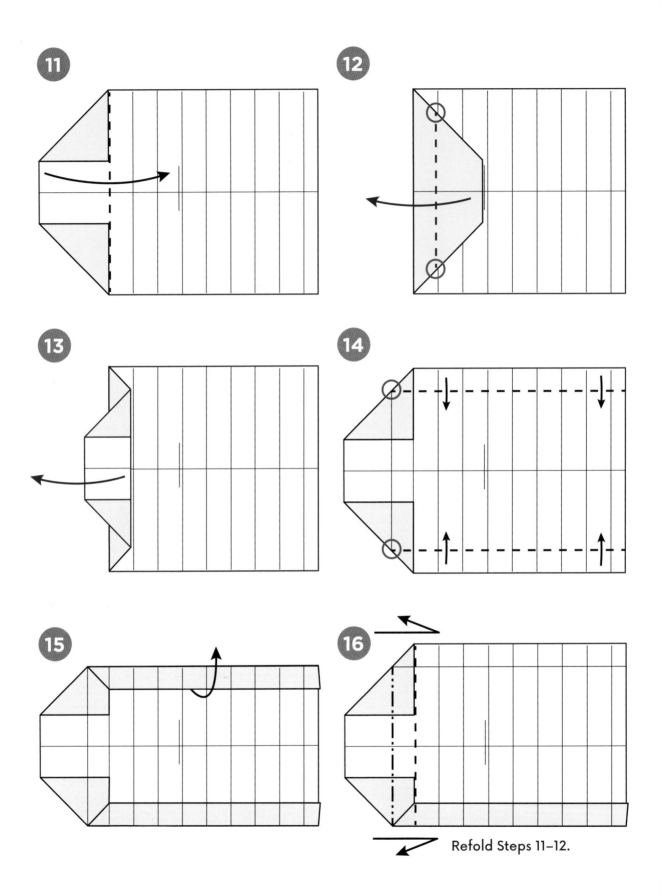

Refold Steps 11–12.

17

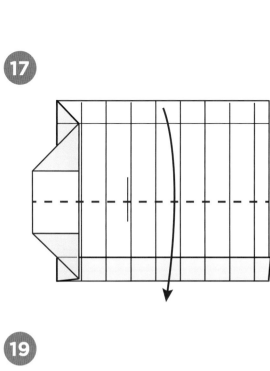

18

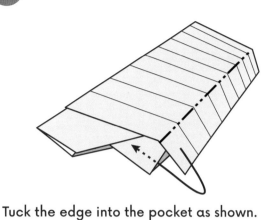

Tuck the edge into the pocket as shown.

19

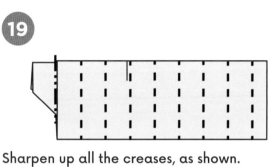

Sharpen up all the creases, as shown.

20

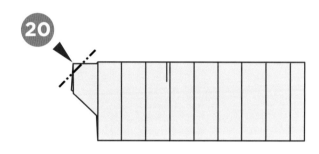

21

Tuck the yellow rectangle into the first rectangle on the left.

22

The mug is an octagonal (eight-sided) shape. Adjust the eight folds and the handle, so that the mug is symmetrical.

23

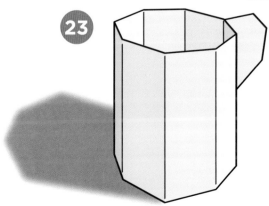

Candle

I like origami candles and have created many: one-piece, multi-piece, 2-D, 3-D ... and more. This four-piece version is very simple to fold, stands well, features a two-color flame (a big "Wow!" factor) and it can be made in a wide variety of decorative papers for extra effect. Can you design another base to support the candle? There must be hundreds of possibilities!

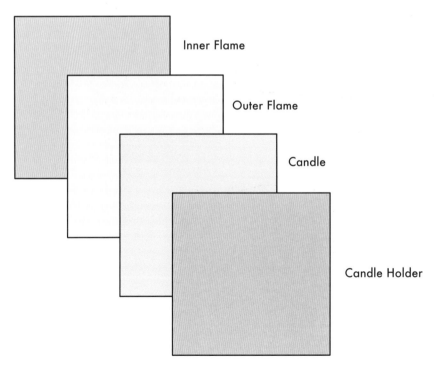

Inner Flame

Outer Flame

Candle

Candle Holder

Candle

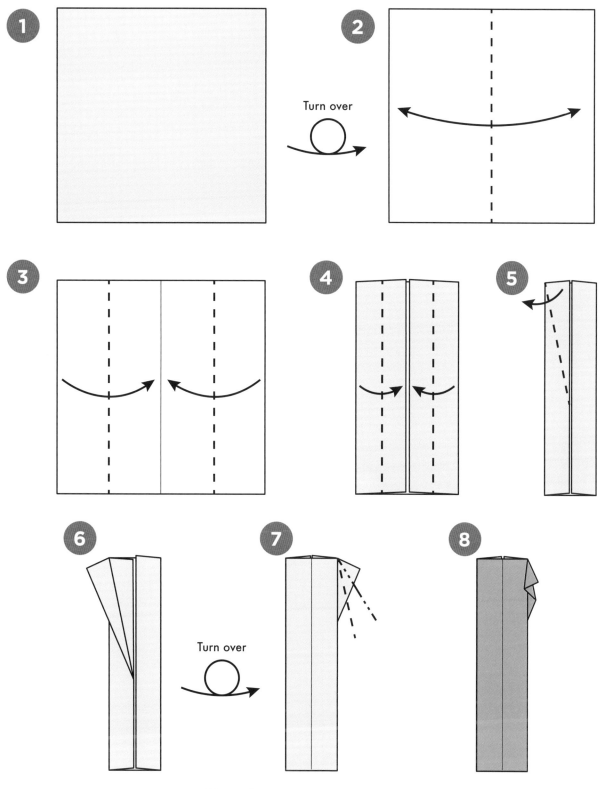

Pleat the paper to represent
a dribble of melted wax.

Candle Holder

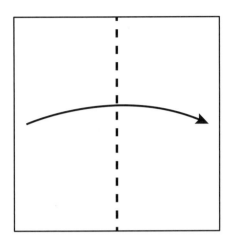

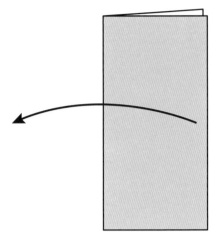

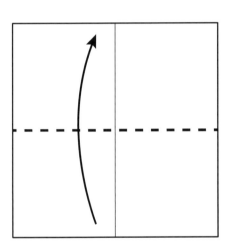

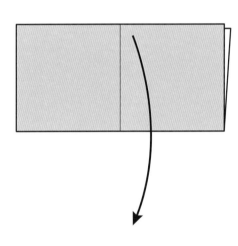

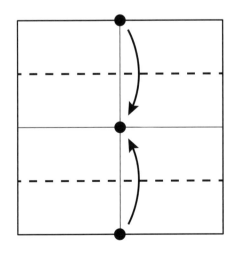

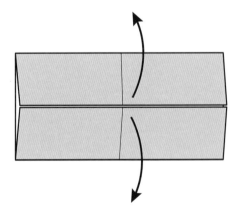

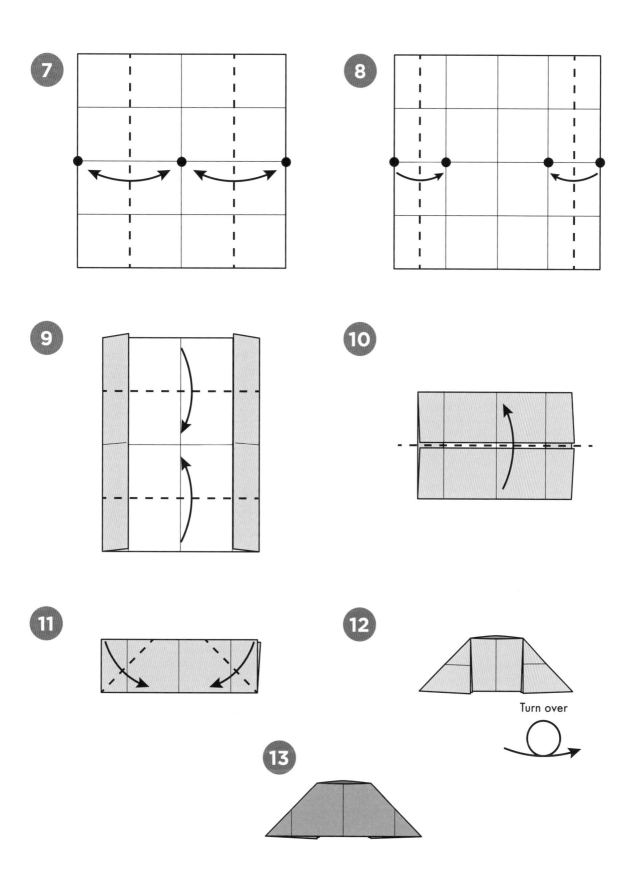

Turn over

Inner and Outer Flames

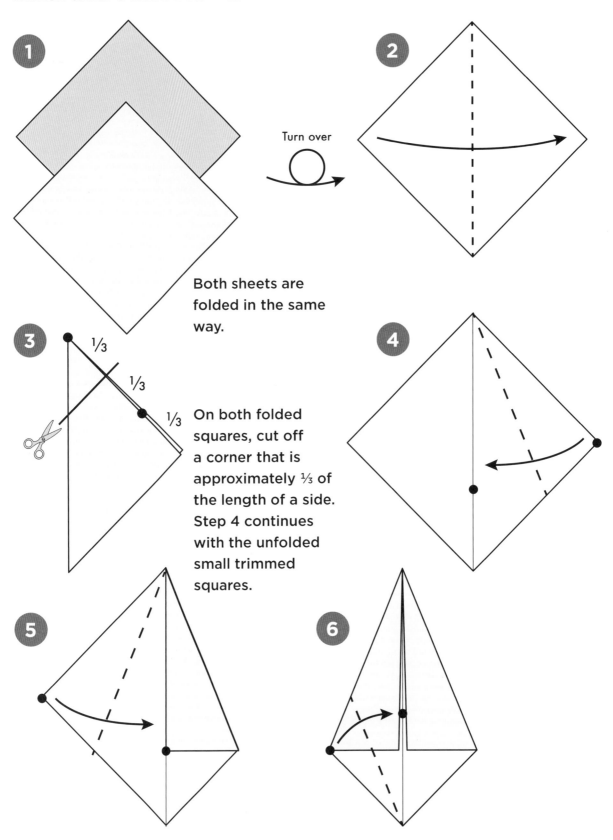

1

2 Turn over

Both sheets are folded in the same way.

3 ⅓ ⅓ ⅓

On both folded squares, cut off a corner that is approximately ⅓ of the length of a side. Step 4 continues with the unfolded small trimmed squares.

4

5

6

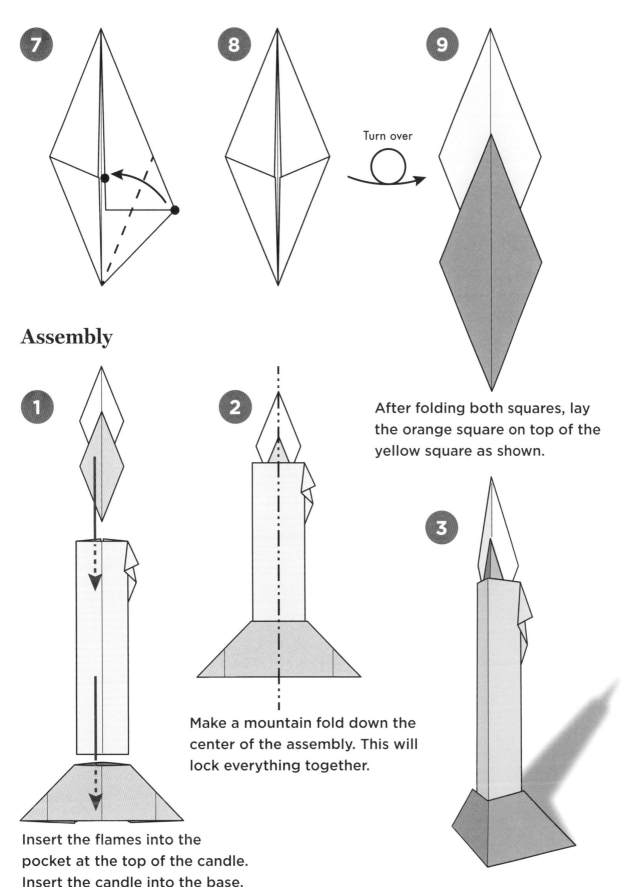

7

8

Turn over

9

Assembly

1

2

After folding both squares, lay the orange square on top of the yellow square as shown.

Make a mountain fold down the center of the assembly. This will lock everything together.

3

Insert the flames into the pocket at the top of the candle. Insert the candle into the base.

Picture Frame

There are many origami picture frames, almost all of which feature small triangles at the corners, into which a picture can be locked. How these triangles are engineered is the first challenge, after which, supporting the frame at an almost vertical angle, becomes the second. Here is my own solution to the problem. Use two squares.

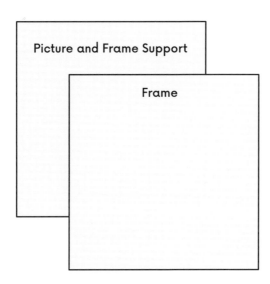

Picture and Frame Support

Frame

Frame

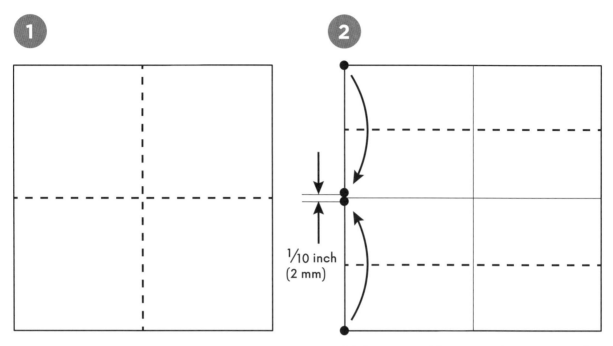

Fold the top and bottom edges toward the center line, leaving a gap of about 1/10 inch (2 mm) between the edges.

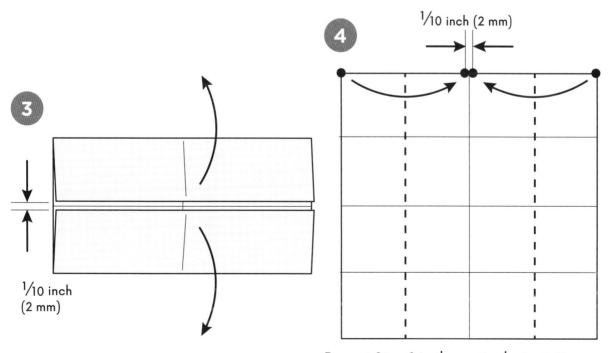

Repeat Step 2 in the vertical orientation.

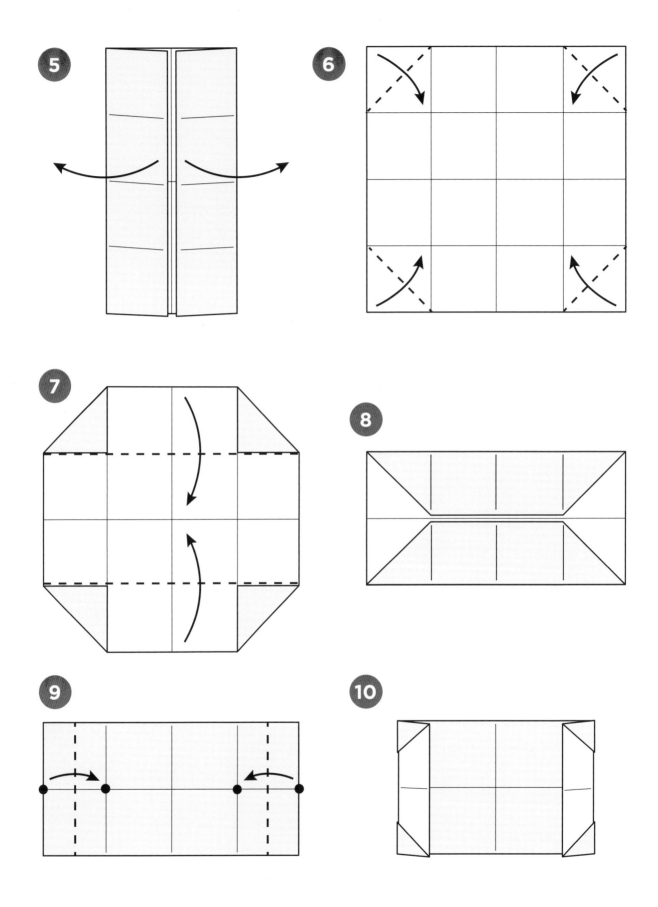

Picture and Frame Support

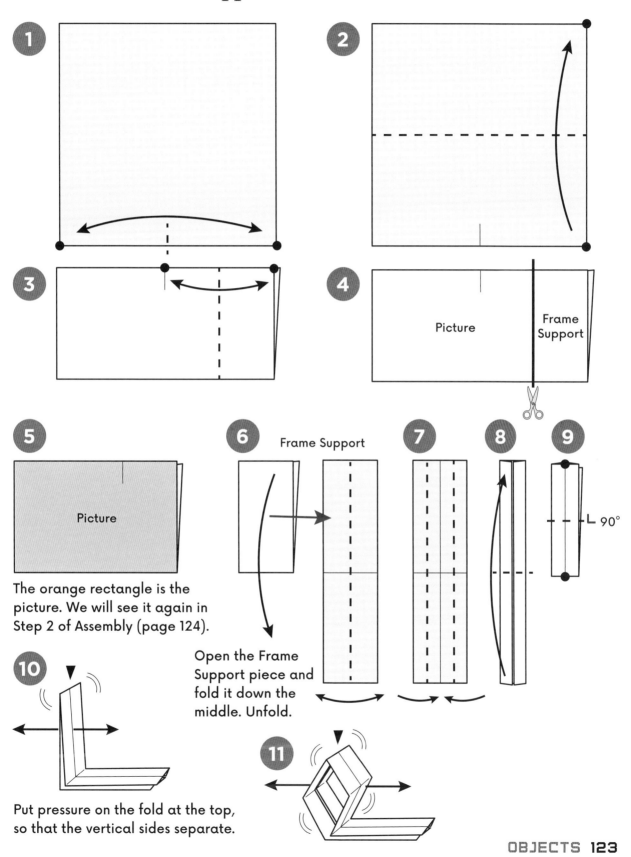

The orange rectangle is the picture. We will see it again in Step 2 of Assembly (page 124).

Open the Frame Support piece and fold it down the middle. Unfold.

Put pressure on the fold at the top, so that the vertical sides separate.

Assembly

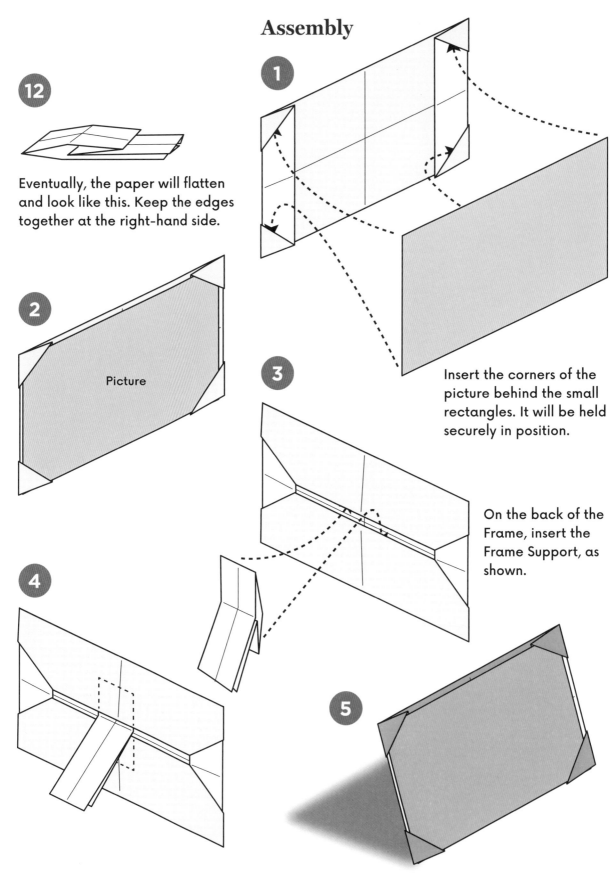

12

Eventually, the paper will flatten and look like this. Keep the edges together at the right-hand side.

1

2

Picture

3

Insert the corners of the picture behind the small rectangles. It will be held securely in position.

On the back of the Frame, insert the Frame Support, as shown.

4

5

Strong Envelope

There are many, many origami envelopes, bags and pouches. However, almost all of them explode when subjected to rough handling. To paraphrase what one cynic once remarked; being practically useless, they are not the real deal, just models of the real deal! By contrast, this design will remain intact and will tear before unfolding itself. It is very practical. Someone suggested that because the bottom corners are missing, it should more correctly be called a "pouch."

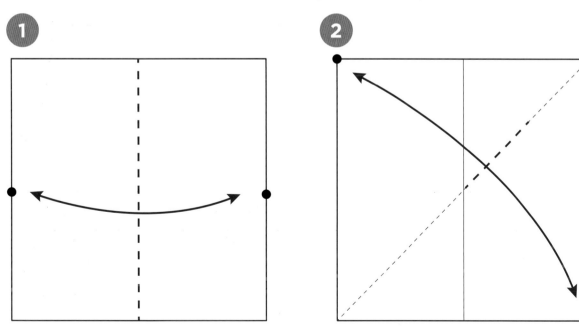

Steps 2–5 consist of a sequence of precise pinched landmarks. Work carefully.

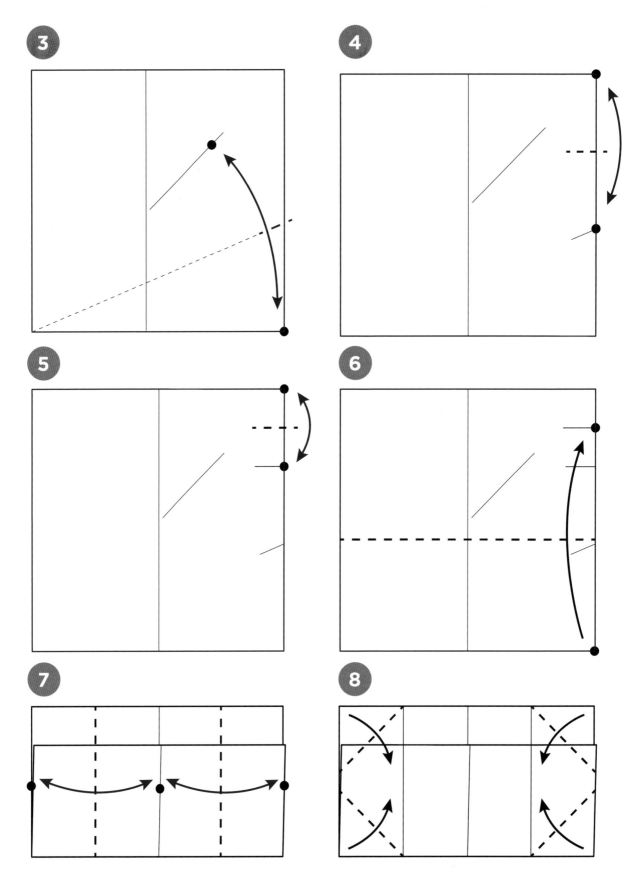

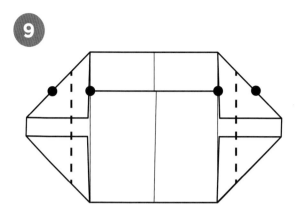

9

Fold dot to dot, as shown. Unfold.

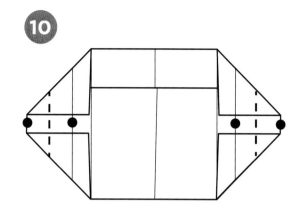

10

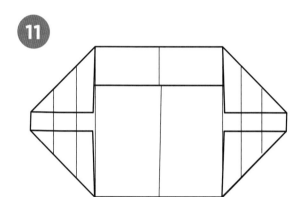

11

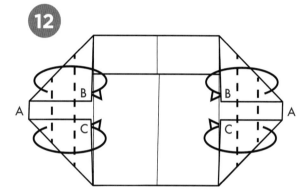

12

Make two folds to tuck A behind B and C. Mirror on the left and right.

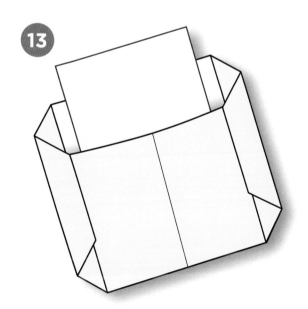

13

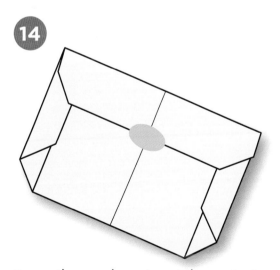

14

To use the envelope, insert the contents, and then fold the top edge down and secure it with a decorative sticker.

Acknowledgments

Most of the designs in this book were created during the COVID lockdowns and taught to different origami groups during the online teaching sessions that proliferated during those difficult weeks and months. For many, these local, national and international online meetings were a chance to chat, to share, to create and to make new friends, when contact with our own social circles was often very limited. For me, teaching at these meetings became a link to the outside world. I thank the organizers of these meetings in the USA, Europe, Africa and Asia for their support of my work, and I thank the many origami enthusiasts whose patience and kindness when folding my work gave me encouragement to create.

I must also thank Ayali and Lio for permitting me to photograph them, Amy Golan-Aharonov for the drawing used in the Picture Frame (depicting an archaeological site near her home) and Kevin Golan-Aharonov for permission to publish his Boat.

Finally, I must also thank my wife, the origami artist and educator, Miri Golan. Her insights into how origami instructions should be drawn and paced to make them as easy as possible to understand formed the basis of the book's visual rhetoric.

Photo & Illustration Credits

p. 2 (middle) MariaNovi / Shutterstock; p. 5 (background) AleksandraMedvedeva / Pixta; pp. 13, 18, 30, 37, 44, 48, 57, 70, 95, 103, 110 (backgrounds) anmbph / Shutterstock; pp. 40, 76, 88 (backgrounds) aga7ta / Shutterstock; pp. 10, 16, 22, 27, 52, 62, 66, 82, 86, 98, 114, 120, 125 (backgrounds) yoshi0511 / Shutterstock; pp. 60, 73, 79, 92, 107 (backgrounds) art_of_sun / Shutterstock

"Books to Span the East and West"

Tuttle Publishing was founded in 1832 in the small New England town of Rutland, Vermont [USA]. Our core values remain as strong today as they were then—to publish best-in-class books which bring people together one page at a time. In 1948, we established a publishing outpost in Japan—and Tuttle is now a leader in publishing English-language books about the arts, languages and cultures of Asia. The world has become a much smaller place today and Asia's economic and cultural influence has grown. Yet the need for meaningful dialogue and information about this diverse region has never been greater. Over the past seven decades, Tuttle has published thousands of books on subjects ranging from martial arts and paper crafts to language learning and literature— and our talented authors, illustrators, designers and photographers have won many prestigious award. We welcome you to explore the wealth of information available on Asia at **www.tuttlepublishing.com**.

Published by Tuttle Publishing, an imprint of Periplus Editions (HK) Ltd.

www.tuttlepublishing.com

Copyright © 2025 Paul Jackson

978-4-8053-1899-7

First edition
29 28 27 26 25   10 9 8 7 6 5 4 3 2 1
Printed in China   2501EP

TUTTLE PUBLISHING® is a registered trademark of Tuttle Publishing, a division of Periplus Editions (HK) Ltd.

Distributed by:

North America, Latin America & Europe
Tuttle Publishing
364 Innovation Drive
North Clarendon,
VT 05759-9436 U.S.A.
Tel: (802) 773-8930
Fax: (802) 773-6993
info@tuttlepublishing.com
www.tuttlepublishing.com

Japan
Tuttle Publishing
Yaekari Building, 3rd Floor
5-4-12 Osaki
Shinagawa-ku
Tokyo 141-0032
Tel: (81) 3 5437-0171
Fax: (81) 3 5437-0755
sales@tuttle.co.jp
www.tuttle.co.jp

Asia Pacific
Berkeley Books Pte. Ltd.
3 Kallang Sector #04-01
Singapore 349278
Tel: (65) 6741-2178
Fax: (65) 6741-2179
inquiries@periplus.com.sg
www.tuttlepublishing.com